Lecture Notes in Mathematics

1937

Editors:
J.-M. Morel, Cachan
F. Takens, Groningen
B. Teissier, Paris

Gianfranco Capriz · Pasquale Giovine
Paolo Maria Mariano (Eds.)

Mathematical Models
of Granular Matter

With Contributions by:

A. Barrat · A.V. Bobylev · C. Cercignani · I.M. Gamba
R. Garcia-Rojo · J.D. Goddard · H.J. Herrmann
S. McNamara · A. Puglisi · T. Ruggeri · G. Toscani · E. Trizac
P. Visco · F. van Wijland

 Springer

Editors

Gianfranco Capriz
Department of Mathematics
University of Pisa
largo Bruno Pontecorvo 5
56127 Pisa
Italy
and
Accademia Nazionale dei Lincei
via della Lungara 10
00165 Roma
Italy
gianfranco.capriz@mac.com

Paolo Maria Mariano
DICeA
University of Florence
via Santa Marta 3
50139 Firenze
Italy
paolo.mariano@unifi.it

Pasquale Giovine
Dipartimento di Meccanica e Materiali
Università "Mediterranea" di Reggio
Calabria
via Graziella, località Feo di Vito
89060 Reggio Calabria
Italy
giovine@unirc.it

ISBN 978-3-540-78276-6   e-ISBN 978-3-540-78277-3
DOI 10.1007/978-3-540-78277-3

Lecture Notes in Mathematics ISSN print edition: 0075-8434
                    ISSN electronic edition: 1617-9692

Library of Congress Control Number: 2008921365

Mathematics Subject Classification (2000): 76T25, 74E20, 74E25, 82D30

Cover design: WMXDesign GmbH

Coverart "Without title", xilo-picture by Luigi Mariano, 1973, private collection

Printed on acid-free paper

9 8 7 6 5 4 3 2 1

springer.com

Preface

The adjective 'granular' is attributed to materials when they are made of sets of unfastened discrete solid particles (granules) of a size larger than one micron, a length scale above which thermal agitation is negligible. In fact, the dominant energy scale in granular materials is the one of a single grain under gravity. Granular matter is common and we meet it everyday. Examples range from the dust settled on the books of our libraries, to the sand in the desert, to the meal used in cooking, itself obtained from grain, often stored in silos. Granular matter displays a variety of peculiarities that distinguish it from other substances studied by condensed matter physics and renders its overall mathematical modeling arduous. In a review paper of 1999 [dG] P.G. de Gennes writes: "granular matter is a new type of condensed matter, as fundamental as a liquid or a solid and showing in fact two states: one fluidlike, one solidlike. But there is as yet no consensus on the description of these two states!"

Almost all preconceptions on which the standard theory of continua is based seem to fail. The standard concept that the material element is well identified (even in the statistical conception common in gas dynamics) fails and, with it, the current mathematical picture assigning to it a precise placement. Even useful results of the standard kinetic theory of gases can be called upon confidently, in general. The populations of grains are far less profuse than the molecular ones in gases and far more crowded. The constraints imposed by grains on one another are generally too conspicuous to rigidify the lot. Also, boundary conditions are far from the simple classical scheme suggested by the divergence theorem and need separate critical modeling.

Heaps of granules do not sustain tension unless (at least a small) cohesion is present. They are, in general, in anisotropic metastable states. Such states last indefinitely unless external perturbations occur. Contrary to common solids and fluids, no thermal average among nearby states arises (see [JNB]). Interactions between granules are exerted through contacts occurring along graphs with topology depending on the way in which granules are packaged, on the distribution of the sizes of the granules themselves, on the boundary conditions and, above all, on the sources of external disturbances. Subsets of granules may self-organize in order to sustain and distribute tensions along arcs: When granules are stored in a silo the pressure at the bottom of the silo does not increase indefinitely as the height of the stored material grows, rather it reaches a maximum value if the lateral walls of the silo are sufficiently tall. If the silo is shaken and grains with different sizes are stored within it, size segregation occurs instead of mixing. The phenomenon does not fit the entropic effects that one recognizes in standard liquids. Both traveling and standing waves accrue in the surface layer of grains, and slip appears along the walls of the silo. It is not clear yet how inelastic interactions and local disorder or segregation contribute to the acoustic propagation.

Layer dynamics is present also in 'avalanches'. The example is the addition of grains to a heap from the top: the surface layer moves, the core persists. The description of the connection between the surface flow and the core at rest forces one to account for the transition between two phases, if one wants to propose a global picture of the phenomenon. As for the contact stresses in granular materials at rest, even for rapid flows the stresses induced by the collisions depend on the local numerosity of granules, in other words on the 'degree of clustering'. Segregation in dense granular flows also has an influence. The overall mechanical behavior seems to be history-dependent.

Plastic effects may be prominent in the quasi-static regime: shear bands appear and inelastic deformations may be accumulated by cyclic loading processes. Inelastic collisions usually play a decisive role in dynamics, as in the fall of avalanches and in the walk of desert dunes. Chaotic agitation of granules leads them far from thermodynamic equilibrium, so fluctuations may be prominent. Microscopic slip friction between granules induces relaxation analogous to that of some solids with complex microstructure. Macroscopic friction is induced by earthquakes which may induce fluidization of granular matter.

Critical reviews such as [JNB], [dG], [K], and [AT] provide an adequate description of phenomena occurring in granular matter. They give a picture of the scenario.

A typical approach to the dynamics of polydisperse granular systems is based on the 'inelastic' Boltzmann equation which also provides the starting point for a plethora of hydrodynamic models. Assumptions on the types of contacts occurring must be chosen carefully: relative rotations may need to be accounted for, depending on the circumstances. Closures are obtained by assuming specific forms of the distribution function. A critical review of the results along these lines can be found in [V].

Even in a disperse state, a granular flow is dissipative as a consequence of the presence of inelastic collisions. Absence of equilibrium is the source of difficulties when one applies the Chapman-Enskog perturbative method to derive a (hydrodynamic) continuum picture from the kinetic description based on Boltzmann equation. Since collisions are inelastic, a double expansion in Knudsen number K and degree of inelasticity ϵ has to be used. Moreover, since

$\epsilon \propto K^3$ in the steady state, the expansion has to be extended up to Burnett or super-Burnett regimes to assure consistency, the latter regime being of degree $O\left(K^3\right)$. The appearance of unphysical instabilities under short-wave perturbations, typical of Burnett and super-Burnett regimes, requires viscosity regularizations of the field equations or the use of other possible techniques.

Attempts have been made to construct continuum models of granular matter from first principles, without resorting to kinetic justifications based on Boltzmann equation, especially in statics. Non-trivial difficulties arise, as already mentioned: at a gross scale some standard paradigms of traditional continuum mechanics need to be modified accurately. We have already mentioned for example the loss of the possibility of identifying perfectly a generic material element (i.e. even to define it). This difficulty is generated by the occurrence of segregation and also by the general lack of coherence due to the absence of cohesion between neighboring granules. Each material element must be then considered as an open system (as in [M]) from which granules may migrate from it to neighboring ones. Standard balances of interactions have to be then supplemented by an equation ruling the rate of local numerosity of granules. Such a rate, which is the rate of migration of granules, generates loss of information about the local texture of the granular matter, and increases the configurational entropy, although segregation is opposite to the entropic mixing (as mentioned above).

The definition of measures of deformation and stresses in terms of granular geometries and grain-to-grain interactions may be non-trivial. As regards the standard stress tensor, for example, a typical definition is made by the sum of the tensor product between the intergranular force and the vector indicating in a local frame the contact point between neighboring granules, a sum extended to all granules in contact with a given granule. However, although such a definition has a physical meaning, one does not know point by point (or better, element by element) the local distribution of granules, their geometry and the type of contact which may be inelastic and not even punctual. All these information characterize each model that can be constructed and have also constitutive nature.

By looking at deeper details, it is natural to consider a granular material as a complex body (in the sense of a body in which the material texture influences strongly the gross behavior). So, multifield descriptions of granular matter need to be called upon, as explained in some chapters of this book.

In any case, no accepted general consensus about the overall description of granular bodies exists. Such an absence of a unified description of the mechanics of the granular matter and the lack of preference for one or another approach, for aesthetic or experimental reasons, has pushed us to collect advances in the various prominent directions of research. Our aim is to furnish a panorama clarifying the state of current researches, solving problems, discussing critically points of view and opening new questions. Contributions range, in fact, from the kinetic approaches to granular flows to the continuum description of static and quasi-static behaviors. A non-trivial tentative of a connection between the kinetic approach and a continuum modeling based on first principles is also present (see Chap. 4).

At the present state of knowledge and in the absence of a unitary point of view, one can only say that the variety and the peculiarities of behavior of granular matter render arduous the task of its overall mathematical modeling. Mathematical and physical questions of an intricate nature appear and tools additional to the traditional ones are needed. Some of these questions and tools are discussed in the subsequent chapters.

Motivated by experimental results on shear bands due to (unstable) plastic behavior of granular bodies, in Chap. 1, *Joe D. Goddard* discusses critically various techniques for the homogenization of granular media in a quasi-static regime, media that are seen here as multipolar continua in the sense of the mechanics of complex bodies. An energy-based method of homogenization is proposed as an improvement on previous approaches.

If one describes granular flows from the point of view of the kinetic theory, inelastic collisions play a role as mentioned above. The Maxwell model of binary collisions is a typical scheme adopted and is based on the assumption that the collision frequency is independent of the velocity of colliding particles. Such a model can be translated from rarefied gases to the description of granular flows. In Chap. 2, *Alexander V. Bobylev, Carlo Cercignani* and *Irene Martinez Gamba* discuss, from a general point of view, variants of the Maxwell model of pairwise interactions and establish their key properties that lead to self-similar asymptotics. Existence and uniqueness issues are also analyzed.

The approach based on the dissipative Boltzmann equation is further analyzed in Chap. 3 by *Giuseppe Toscani*. Two paths toward the hydrodynamic limit are discussed. They account for two different methodologies for the closure of macroscopic equations: (i) low inelasticity in the system, namely a perturbation in a precise sense that allows the local resort to Maxwellian functions, (ii) small spatial variations implying the use of a homogeneous cooling state.

One of the editors (**GC**) proposes in Chap. 4 (a chapter already mentioned above) further results on an earlier proposal for fast sparse flows. Complex bodies with kinetic substructure are considered. They are bodies in which each material element is a system in continuous agitation and are called *pseudofluids*. Granular flows fall naturally in this framework. The maelstrom within each material element and its influence on the neighboring fellows is governed by peculiar hydrodynamic balances: they are offspring of the microscopic interactions between granules.

The recursive analysis of higher momenta of the distribution function, even beyond Grad's 13 moments, has been the basic source of Extended Thermodynamics. The results on monoatomic gases provided by such an approach suggest its use in modeling granular flows. After summarizing the main modeling issues of Extended Thermodynamics, in Chap. 5 *Tommaso Ruggeri* analyzes the resulting hyperbolic system and discusses global existence questions and stability of constant state on the basis of Kawashima condition.

The construction of hydrodynamic models from first principles can find appropriate suggestions from detailed numerical simulations performed by looking directly at single granules and their contact interactions. The molecular dynamic approach is pursued in Chap. 6 by *Ramon García-Rojo*, *Sean McNamara* and *Hans J. Herrmann* who analyze (amid possible choices) the persistent elastic–plastic strain accumulation in compacted granular soils under cyclic stress conditions.

Since driven sets of granules (for example confined in a box) are in essence systems very far from thermodynamic equilibrium, as mentioned above, the effects of fluctuations are in general significant. Analysis of the injected power fluctuations is presented by *Alain Barrat*, *Andrea Puglisi*, *Emmanuel Trizac*, *Paolo Visco* and *Frederic van Wijland* in Chap. 7, a chapter divided into two parts: The first part deals with the way in which the probability density function of the fluctuations of the total energy is related to the characterization of energy correlations for both boundary and homogeneous driving. The second part contains an interpretation of some numerical and experimental results that seem to invalidate Gallavotti–Cohen symmetry [GC]. Such results appear contradictory to common analyses that seem to satisfy Gallavotti–Cohen fluctuation relation. By means of Lebowitz–Spohn approach to Markov processes, an approach applied to the inelastic Boltzmann equation, a functional satisfying a fluctuation relation is also introduced.

The last two editors collect their contributions in Chaps. 8 and 9. In particular, in Chap. 8 **PG** analyzes in the continuum limit the thermodynamics of a granular material modeled as a complex body endowed with a microstructure which is constrained and/or latent in the sense introduced by Capriz [C]. The consequences of grain rotations are described together with effects like dilatancy.

Finally, in Chap. 9, **PMM** considers granular matter in slow motion and describes it as a two-scale complex body for which each material element is considered as a grand-canonical ensemble of granules. The evolution equation of the numerosity of grains in each material element is derived in terms of grain-to-grain interactions.

February, 2008

Gianfranco Capriz
Pasquale Giovine
Paolo Maria Mariano

References

[AT]   Aranson, I.S., Tsimring, L.S.: Patterns and collective behavior in granular media: Theoretical concepts. Rev. Modern Phys., **78**, 641–692 (2006)

[C]   Capriz, G.: Continua with latent microstructure. Arch. Rational Mech. Anal., **90**, 43–56 (1985)

[dG]   de Gennes, P.G.: Granular matter, a tentative view. Rev. Modern Phys., **71**, S375–S382 (1999)

[G]   Goldhirsch, J.: Kinetic and continuum descriptions of granular flows. In: Aref, H., Philips, J.W. (eds) Mechanics for a New Millenium. Philips Edts., Kluwer, Dordrecht, 345–358 (2001)

[GC]   Gallavotti, G., Cohen, E.G.D.: Dynamical ensembles in nonequilibrium statistical mechanics. Phys. Rev. Letters, **74**, 2694–2698 (1995)

[K]   Kadanoff, L.P.: Built upon sand: Theoretical ideas inspired by granular flows. Rev. Modern Phys., **71**, 435–444 (1999)

[JNB]   Jaeger, H.M., Nagel, S.R., Behringer, R.P.: Granular solids, liquids, and gases. Rev. Modern Phys., **68**, 1259–1273 (1996)

[M]   Mariano, P.M.: Migration of substructures in complex fluids. J. Phys. A: Math. Teoretical, **38**, 6823–6839 (2005)

[V]   Villani, C.: Mathematics of granular materials. J. Stat. Phys., **124**, 781–822 (2007)

Contents

List of Contributors

Alain Barrat
Laboratoire de Physique Théorique
(CNRS UMR 8627),
Université Paris-Sud
Orsay, France
Alain.Barrat@th.u-psud.fr

Alexander V. Bobylev
Department of Mathematics,
Karlstad University
Universitetsgatan 2,
65188 Karlstad, Sweden
alexander.bobylev@kau.se

Gianfranco Capriz
Department of Mathematics,
University of Pisa
largo Bruno Pontecorvo 5,
56127 Pisa, Italy
gianfranco.capriz@mac.com

Carlo Cercignani
Dipartimento di Matematica,
Politecnico di Milano
via Bonardi 9,
20133 Milano, Italy
carcer@mate.polimi.it

Ramon García-Rojo
Fisica Teorica, Facultad de Fisica,
Universidad de Sevilla
Apartado de Correos 1065,
41080 Sevilla, Spain

Pasquale Giovine
Dipartimento di Meccanica
e Materiali,
Università "Mediterranea"
via Graziella,
Località Feo di Vito,
89127 Reggio Calabria,
Italy
giovine@unirc.it

Joe D. Goddard
Department of Mechanical
and Aerospace Engineering,
University of California
at San Diego - La Jolla
La Jolla, CA 92093-0411, USA
jgoddard@ucsd.edu

Hans J. Herrmann
Departamento de Fisica,
Universidade Federal do Ceará
Campus do Pici,
60455-900 Fortaleza
Brazil
hans@fisica.ufc.br

Paolo Maria Mariano
DICeA, University of Florence
via Santa Marta 3,
50139 Firenze, Italy
paolo.mariano@unifi.it

Irene Martinez Gamba
Department of Mathematics,
The University of Texas
at Austin
Austin, TX 78712, USA
gamba@math.utexas.edu

Sean McNamara
ICP, University of Stuttgart,
Pfaffenwaldring 27,
70569 Stuttgart, Germany
S.McNamara@icp.uni-stuttgart.de

Andrea Puglisi
Dipartimento di Fisica,
Università di Roma
"La Sapienza"
Piazzale Aldo Moro 2,
00185 Roma, Italy
andrea.puglisi@roma1.infn.it

Tommaso Ruggeri
Research Center for Applied
Mathematics, C.I.R.A.M.
University of Bologna
via Saragozza 8,
40123 Bologna, Italy
ruggeri@ciram.unibo.it

Giuseppe Toscani
Department of Mathematics,
University of Pavia
via Ferrata 1,
27100 Pavia, Italy
giuseppe.toscani@unipv.it

Emmanuel Trizac
Laboratoire de Physique,
Théorique et Modèles Statistiques
(CNRS UMR 8626),
Université Paris-Sud
Orsay, France
trizac@ipno.in2p3.fr

Frederic van Wijland
Laboratoire Matière et Systèmes
Complexes (CNRS UMR 7057),
Université Denis Diderot (Paris VII),
Paris, France
Frederic.Van-Wijland
@th.u-psud.fr

Paolo Visco
Laboratoire de Physique,
Théorique et Modèles Statistiques
(CNRS UMR 8626),
Université Paris-Sud
Orsay, France
Paolo.Visco@th.u-psud.fr

From Granular Matter to Generalized Continuum

J.D. Goddard

Summary. Following a cursory review and synthesis of multipolar continua, the rudiments of graph theory, and granular mechanics, a graph-theoretic, energy-based homogenization is proposed for the systematic derivation of multipolar stress and kinematics in granular media. This provides a weakly non-local hierarchy of multipolar field equations for quasi-static mechanics based on polynomial representations of the kinematics of the type employed in past works. As an improvement on those works, a method is proposed for avoiding "overfitting" of fluctuations based on the so-called "Generalized Additive Method" of statistics. Among other results, it is shown that the standard formula for Cauchy stress in granular media may break down owing to multipolar effects, and that granular rotations in the typical granular medium should not lead to Cosserat effects, as the lowest-order departure from the simple-continuum model.

1 Introduction

Arguments against new ideas generally pass through distinct stages from:

> "It's not true" to
> "Well it's true but not important" to
> "It's true and it's important, but it's not new – we knew it all along"

> From *The Artful Universe*
> by John D. Barron
> (Chap. 1 of [16])

This article, an amended and enlarged version of a recent conference paper [21], has its beginnings in a much older work [18] concerned with the largely theoretical question as to the definition of stress in a granular assembly. By no means novel at the time, the question has taken on a more practical importance in the intervening years, in part motivated by "shear bands" associated with the unstable plasticity of granular media.

According to simple continuum models, shear bands represent infinitely thin surfaces of discontinuity, in distinct contrast to the zones of finite thickness revealed by numerous experiments and particle-level computer simulations. Moreover, certain of these studies indicate that the grain rotation in shear bands may be quite different from the global rotation (vorticity) outside the band, which is generally interpreted as a manifestation of "Cosserat" effects.

As anticipated in the general field of plasticity and soil mechanics, some type of "enriched" or "structured" continuum model endowed with intrinsic length scale is required to regularize the underlying field equations in the presence of material instability. Furthermore, the additional forces implicated in such models may actually influence the onset and evolution of material instability, as recognized early on by Vardoulakis and coworkers [36, 46].

Given the overall progress of granular mechanics in the last two decades, a renewed effort to elucidate the above theoretical questions seems timely and appropriate. With this motivation, the present paper provides a synthesis and critique of various principles and techniques for the homogenization of granular media, with emphasis on the quasi-static mechanical behavior.

A brief review is presented of multipolar continua, regarded as general models for granular media, and a survey is given of the graph-theoretic and energy principles underlying granular micromechanics, based on the interpretation of the associated matrices as differential operators. A novel energy-based method is proposed for homogenization, as a modification of the abstract "best fits" proposed elsewhere. This method employs polynomial representations for particle displacements and forces which provide the relevant gradients and moment stresses for micropolar continua. Based on the works of Eringen [16], a general formula is postulated to include contributions from the motion of particle centroids, from particle deformation, and from singular surfaces exhibiting slip or interfacial tension.

As new results, it is shown that:

1. In the absence of intergranular contact moments or external body couples, grain rotations do not contribute directly to the quasi-static stress work, in particular to frictional dissipation, and, therefore,
2. The resultant Cauchy stress derives solely from the motion of particle centroids.

A major goal of the present work is to establish the micropolar continuum as a plausible model for granular and cellular media, by showing in a general and systematic way how micropolar effects emerge from discrete micromechanical models. A second goal is to present a concise formulation of the underlying mathematical techniques, and to connect to basic ideas from topology, graph theory and to other fields of network analysis.

1.1 Mathematical Preliminaries

The notation is similar to that employed in a previous work [19], where bold symbols are employed for space tensors, lowercase symbols for vectors $\varphi = \varphi^\alpha \mathbf{g}_\alpha$, uppercase for higher-order tensors $\mathbf{L} = L^{\alpha\beta\cdots} \mathbf{g}_\alpha \otimes \mathbf{g}_\beta \otimes \ldots$, etc., where $\otimes$ denotes a tensor product, and Greek superscripts and subscripts refer to a basis $\mathbf{g}_\alpha, \alpha = 1, 2, 3$, derived from appropriate spatial coordinates. For the present purposes, the latter may be taken as orthogonal cartesian. A colon is employed denote the exhaustive, ordered contractions of tensors of rank two and higher, such that, for $n \geq m$, $(L^{\alpha_1\cdots\alpha_n}) : (M^{\beta_1\cdots\beta_m}) := L^{\alpha_1\cdots\beta_1,\cdots\beta_m} M_{\beta_1\ldots\beta_m}$, and we employ superscript T to denote transposition of the right-most tensor component with all the preceding, so that $(L^{\alpha_1\alpha_2\cdots\alpha_n}\ldots)^T := (L^{\alpha_n\alpha_1\alpha_2\cdots}\ldots)$. The standard notation $\mathbf{Lx}(=L^\alpha_\beta x^\beta \mathbf{g}_\alpha)$ is employed for linear transformations of vectors via second-rank tensors. Roman superscripts are used throughout (in contrast to [21]) to label particles (i.e. grains), branch vectors and the associated graph-theoretical matrices in granular assemblies. Brackets [,] are employed to denote closed intervals of both reals and integers, and the standard symbol \ denotes set exclusion.

With $\mathbf{a}^n$ denoting the n-fold symmetric tensor product $\otimes^n \mathbf{a}$, the Taylor series expansion for the velocity (or infinitesimal displacement) $\mathbf{v}$,

$$\mathbf{v}(\mathbf{x}) = \mathbf{v}_o + \mathbf{L}_1 \mathbf{r} + \mathbf{L}_2 : \mathbf{r}^2 + \ldots, \tag{1}$$

with

$$\mathbf{r} = \mathbf{x} - \mathbf{x}_o, \quad \mathbf{L}_n = \frac{1}{n!} (\boldsymbol{\nabla}^n \otimes \mathbf{v})_o^T, \tag{2}$$

provides the well-known expansion for global stress-power density in a simple continuum

$$\dot{w} = \frac{1}{V} \int_V \mathbf{T} : \mathbf{L} \, dV = \sum_n \dot{w}_n, \tag{3}$$

with

$$\dot{w}_n := \mathbf{T}_n : \mathbf{L}_n, \quad \mathbf{T}_n := \int_V \mathbf{T} \otimes \mathbf{r}^n dV, \tag{4}$$

where $\mathbf{T}$ is Cauchy stress and $\mathbf{L} = (\boldsymbol{\nabla} \otimes \mathbf{v})^T$ (first) velocity gradient. With stress moments $\mathbf{T}_n$ representing generalized forces conjugate to the kinematical quantities $\mathbf{L}_n$ [19, 25]. Equation (4), serves to establish an equivalence between a non-homogeneous simple continuum and a homogeneous *multipolar* continuum, i.e. a continuum endowed with intrinsic moment stresses. We distinguish two important special cases:

1. The $\mathbf{L}_n$ are identical with higher gradients of the local velocity field, as defined in (2), or
2. They are intrinsic "particulate" fields, say $\mathbf{L}_n^p(\mathbf{x}, t)$, given by more general constitutive equations.

The first represents a *graded* (or "Toupin–Mindlin") continuum [23, 25, 35], while the second represents a *micromorphic* (or "Cosserat–Eringen") continuum [14, 24, 35, 41].[1] Both are endowed with intrinsic length scales, and the graded continuum can be regarded as a manifestation of weak non-locality, a precursor to a fully non-local continuum [17].

By means of the mathematical "fragmentation" of a simple continuum into discontinuous subdomains, Eringen and coworkers have derived micromorphic field theories resembling those obtained by certain statistical mechanical studies of systems of deformable particles [16]. A similar technique has been presented recently for the special case of Cosserat media [13]. The micromorphic continuum is a special case of a multipolar continuum endowed with a *polyad* of deformable vectors or "directors" attached to each material particle [24], with $\mathbf{L}_n^p$ representing 3^n such vectors.[2] The simplest ("grade one") micromorphic continuum is characterized by deformable triad of vectors and, hence, a single second-rank (velocity gradient) tensor $\mathbf{L}^p$ attached to each material point that serves to represent an homogeneous microstructural ("particle") deformation and rotation. In the special case of a micropolar (Cosserat) continuum, $\mathbf{L}^p = \mathbf{W}^p = -(\mathbf{W}^p)^T$ and $\boldsymbol{\omega}^p = \text{vec}(\mathbf{W}^p)$ represent a (particle) spin generally distinct from the global spin $\mathbf{W} = (\mathbf{L} - \mathbf{L}^T)/2$. (Recall that [16] distinguishes micropolar as the rigid subclass of microstretch, the isotropic or spherical subclass of micromorphic.)

The various moment stresses may be interpreted, as above, in terms of volumetric working, or alternatively, in terms of their infinitesimal surface actions $\mathbf{T}_n \cdot d\mathbf{s}$. Thus, for $n = 1$, one has a force ("push" or "pull"), inducing displacement, and for $n = 2$, symmetric and skew-symmetric moments (generalized "pinch"),[3] inducing stretch and rotation, respectively, etc. For later reference, we define a micropolar continuum of grade m as one having $w_n \equiv 0$ for $n > m$ for all deformation histories, the simple continuum [43] being *grade one*.

1.2 Balances

According to a microscopic treatment [19], the stress moments $\mathbf{T}_n$ in a multipolar continuum should satisfy a hierarchy of balances of the form:

$$\nabla \cdot \mathbf{T}_{n+1}^T + \mathbf{T}_n = \mathbf{G}_{n+1}, \text{ for } n = 0, 1, \ldots, \tag{5}$$

where $\mathbf{T}_0 := \mathbf{0}$, $\mathbf{T}_1 := \mathbf{T}$, and the $\mathbf{G}$'s represent extrinsic body moments plus accumulation of intrinsic multipolar momenta. The latter are not made

---

[1] Both were designated as "micromorphic" in [21], based on the idea that response to higher gradients involves some finite microstructure. Abandoning that unconventional usage, we adopt another, with "graded" designating what is sometime called "gradient" or "higher-gradient" continuum.

[2] The basic idea is attributed to Duhem in [44]. Since directors are attached to material points, we exclude from (3) et seq. a term $\dot{w}_0 = \mathbf{f}_0 \cdot \mathbf{v}_0$ involving a body force $\mathbf{f}_0$ and a relative velocity $\mathbf{v}_0$, appropriate to two-phase media and anticipated by the theory of mixtures [44].

[3] A term suggested by Professor I. Vardoulakis [45].

explicit here, since they are negligible in the quasi-static limit. It is easy to verify that (5) leads to the following integral balance:

$$\int_{\partial V} [\mathbf{T}_n + \mathbf{x} \otimes \mathbf{T}_{n-1} + \ldots + \mathbf{x}^{n-1} \otimes \mathbf{T}] \cdot d\mathbf{s} = \int_V \mathbf{G}_n dV, \text{ for } n = 2, 3, \ldots, \quad (6)$$

in which all stresses up to order n contribute to the nth surface moment.

The uncertain status of (5) should be acknowledged immediately, since it has not been established by any of the standard methods of continuum mechanics, neither by derivation from the energy balance, by means of invariance principles (objectivity) [15, 16, 22], nor from variational principles for elastic systems, dating back to the Cosserats [11, 44] (and treated in a recent review [30] that includes both micromorphic and graded continua). At any rate, the balance for $n = 1$ has the form found elsewhere, up to an arbitrary additive symmetric, second-rank stress [16]. In this author's opinion, the latter might be profitably be regarded as the divergence of a third rank tensor and absorbed into the term $\mathbf{\nabla} \cdot \mathbf{T}_2^T$ in (5).

No attention is paid here to compatibility, discontinuity and boundary conditions for the various kinematic tensors $\mathbf{L}_n$, some of which are discussed in [16]. Also, we do not deal with discontinuity and boundary conditions for the associated moment stresses $\mathbf{T}_n$, since the main focus of this article is on the issues surrounding the passage from discrete microstructure to continuum model.

2 Micromechanics

The kinematics of granular media involves both *extrinsic* modes or degrees of freedom, associated with motion of particle centroids, and *intrinsic* or internal modes associated with particle deformation. Although the two are coupled mechanically through particle contacts, we first focus attention on the extrinsic modes. For the sake of completeness, the following subsection provides some essential background material on granular media [1, 19] and elucidates the role of grain rotation.

2.1 Granular Microstructure and Rotation

Figure 1 illustrates the standard idealization of a granular medium [19, 37], with $i, j \in [1, N]$ enumerating particles or grains. With $\mathbf{x}$ denoting the position vector, dotted lines represent moment arms $\mathbf{r}^{ij} = \mathbf{x}^i - \mathbf{x}^{ij}$ connecting grain centroid $\mathbf{x}^i$ to nominal point of contact $\mathbf{x}^{ij}$, neighbors j being defined by triangulation on centroids (vide infra). Solid and dashed lines then represent branch vectors $\mathbf{l}^{ij} = \mathbf{r}^{ij} - \mathbf{r}^{ji} = -\mathbf{l}^{ji}$, with solid lines indicating active contacts and dashed lines representing virtual contacts (i.e. nearest neighbors without contact).

Fig. 1. Idealized granular medium

The interparticle contact force $\mathbf{f}^{ij}$ is the resultant defined by the surface integrals on the left-hand side of (6) with $n = 1$, taken over a nominal contact area ij. As shown in numerous preceding works, e.g. [1], the assumption of contact forces $\mathbf{f}^{ij} = -\mathbf{f}^{ji}$ localized at points $\mathbf{x}^{ij}$ leads to a particle contribution to the volume-average stress (a "dipole") given by

$$\mathbf{T}^i = \frac{1}{V^i} \sum_j \mathbf{f}^{ij} \otimes \mathbf{r}^{ij}, \tag{7}$$

for each particle i in the interior of the granular assembly.

The vector couple about $\mathbf{x}^i$ exerted by particle j on i is given

$$\mathbf{c}^{ij} = \mathbf{f}^{ij} \times \mathbf{r}^{ij} + \mathbf{m}^{ij}, \tag{8}$$

where $\mathbf{m}^{ij} = -\mathbf{m}^{ji}$ is the vector of the skew part of an integral of the type (6), with $n = 2$, with $\mathbf{x}$ replaced by $\mathbf{x} - \mathbf{x}^i$, and with $\mathbf{x}^{ij}$ representing the centroid of the contact stress. Thus, for grains composed of a simple (grade one) material, the contribution from $\mathbf{T}_2$ vanishes, so that the contact couple $\mathbf{m}$ arises solely from the moment $(\mathbf{x} - \mathbf{x}^i) \times \mathbf{T} \cdot d\mathbf{s}$ as a kind of rolling resistance.

In the absence of external body couples and contact moments $\mathbf{m}^{ij}$, the quasi-static moment balance requires the symmetry of $\mathbf{T}^i$. Furthermore, it is easy to show that

$$\sum_i V^i \mathbf{T}^i = \sum_{i>j} \mathbf{f}^{ij} \otimes \mathbf{l}^{ij}, \tag{9}$$

mapping particle dipoles into contact dipoles.

On the other hand, the power of internal contact forces is given by the sum over distinct contacts

$$\dot{W} = \sum_{i>j} \mathbf{f}^{ij} \cdot \mathbf{v}^{ij} \tag{10}$$

with

$$\mathbf{v}^{ij} = \mathbf{u}^{ij} + \mathbf{W}^i \mathbf{r}^{ij} - \mathbf{W}^j \mathbf{r}^{ji}, \text{ with } \mathbf{u}^{ij} = \dot{\mathbf{x}}^i - \dot{\mathbf{x}}^j, \tag{11}$$

where W is the skew-symmetric tensor representing particle rotation. Hence, it follows readily that

$$\dot{W} = \sum_{i>j} \mathbf{f}^{ij} \cdot \mathbf{u}^{ij} + \sum_{i} V^i \mathbf{T}^i : \boldsymbol{W}^i. \tag{12}$$

Since the second term vanishes whenever $\mathbf{T}^i$ is symmetric, we have for arbitrary particle shapes the

Theorem. *In the absence of external body couples and intergranular contact moments* ($\mathbf{m}^{ij}$) *the rotation of internal grains makes no direct contribution to quasi-static stress power.*

Here, "internal" refers to those grains in mechanical contact with other grains but not with any boundary from which couples may be transmitted. Of course, the resultant of the latter must be zero.

The preceding theorem has implications for quasi-static Cosserat effects, since most existing micromechanical predictions of such effects, as typified by [31,33,42], depend explicitly on contact moments. However, since the linear dimension of typical (Hertzian) intergranular contact zones, proportional to some $O(1)$ power of the ratio of confining pressure to elastic modulus, is expected to be small, especially for rigid noncohesive geomaterials such as sand, it follows that the term $\mathbf{f} \times \mathbf{r}$ will dominate the term $\mathbf{m}$ in (8).

The same conclusion results for rigid noncohesive particles with multiple contact zones, since the contact forces on such zones can be replaced by a finite, statically equivalent set of forces, whose moments are once again captured by the terms of the form $\mathbf{f} \times \mathbf{r}$ in (8). Hence, for nearly rigid grains, any homogenization scheme based on energy principles should yield negligible quasi-static Cosserat effects, although particle rotations will generally have other influences on the micromechanics.

It should be noted that certain types of cohesive contacts may give rise to important contact moments, since they allow for a locally large tensile force balanced by a locally large compressive force, producing a large couple $\mathbf{m}$ without a correspondingly large resultant $\mathbf{f}$.

The absence of intergranular contact moments not only justifies various simple-continuum models of granular media but also implies that the breakdown of such models must be due to effects other than those associated with Cosserat rotation. In this connection, we note that the symmetry of (7) and hence of (9), implies that $\mathbf{T}^T = \mathbf{T}$ (Cauchy's second law [43]), at least according to the standard formula for $\mathbf{T}$. However, the latter is subject to criticisms presented below.

2.2 Graph Theory for Extrinsic Modes

Graph theory provides a particularly appealing tool for the description of various physicochemical networks. In contrast to the mechanical networks associated with structural mechanics [40] or statistical physics [5], the graphs for granular media or other mobile cellular assemblies are often transitory, reflecting abrupt topological rearrangement engendered by finite deformation and requiring a sequence of graphs to describe evolving microstructures. Immediately following is a concise treatment of the graph-theoretical description of granular mechanics, with connections to other fields of application and to the basic mathematical literature.

In a schema dating back to the early works of Satake [31, 37, 38],[4] we let particle centroids define the nodes or vertices $j \in [1, N]$ defined by an appropriate Delaunay triangulation [1,19,20,37]. Note that this triangulation should be generally based on (minimal) separation between particle surfaces rather than particle centroids. Whenever unique, this defines an abstract (connected simple) graph $\mathcal{G}$, the *granular contact network* or *Satake graph*, with edges or branches $i \in [1, E]$ representing nominal nearest-neighbors and defining contacts or virtual contacts.

In the associated matrix formulation, underlined lowercase quantities denote columns (vertical arrays) associated with edges and nodes, while superscript $*$ denotes transposition (vector-space dual), e.g. $\varphi = [\varphi^i]^* = [\varphi^1, \ldots, \varphi^N]^*$ denotes a $1 \times N$ row of scalars, $\underline{\varphi} = [\varphi^i]^* = [\varphi^1, \ldots, \varphi^E]^*$, a $1 \times E$ row (horizontal array) of space vectors, etc. Then, underlined uppercase denotes the associated linear transformations or matrices, e.g. $\underline{A} = [A^{ij}]$, $\mathbf{\underline{A}} = [\mathbf{A}^{ij}]$, etc., with dual or adjoint defined by the standard scalar products $(\underline{u}, \underline{v}) = (\underline{v}, \underline{u}) := \underline{u}^* \underline{v}$ and $(\mathbf{\underline{u}}, \mathbf{\underline{v}}) := \mathbf{\underline{u}}^* \cdot \mathbf{\underline{v}} = \sum_k \mathbf{u}^k \cdot \mathbf{v}^k F$.

Assignment of directions to the edges of the above graph yields a directed graph [4,6], with $E \times N$ incidence matrix $\underline{D} = [D^{ij}]$

$$D^{ij} = \begin{cases} +1, \text{if edge } i \text{ enters node } j, \\ -1, \text{if edge } i \text{ leaves node } j, \\ 0, \text{otherwise.} \end{cases} \tag{13}$$

The matrix $\underline{D}$ (the transpose of the matrix D in [5,6,38]) and its transpose $\underline{D}^*$ represent difference-operators, which we designate, respectively, as the *differential* and the *codifferential* (denoted respectively as "coboundary" and "boundary" operators in the standard literature on graph, e.g. p. 5 of [5]). Thus, $\underline{D}\varphi$ yields differences along edges of nodal "potentials" represented by φ, while $\underline{D}^* f$ yields nodal accumulations from flows along edges ([6] and p. 5 of [5]). Since it can be shown that the column rank of $\underline{D}$ is $N-1$, we henceforth delete the final column,[5] corresponding to "ground" node N, and denote the resulting $E \times (N-1)$ matrix by the same symbol $\underline{D}$. Accordingly, we dispense with the Nth component of column operands, reducing them to $(N-1) \times 1$ arrays.

---

[4] The present treatment does not rely explicitly on the geometric properties of voids nor on the composite (Schaefer) operators employed in [38] to express micromorphic compatibility, discussed more generally in [16].

[5] Not usually done in the prevalent literature and not made explicit in [21].

Another important operator $\underline{D}_\times$ is given by

$$\underline{D}^*\underline{D}_\times = \underline{0}. \text{ and } \underline{D}^*_\times\underline{D} = \underline{0}, \text{ i.e. } \underline{D}_\times = \ker(\underline{D}^*), \tag{14}$$

ker denoting the kernel or null space of a linear transformation. The matrix $\underline{D}_\times$ can be taken as any $E \times M$ matrix whose M columns form a basis for the null space of $\underline{D}^*$, where $M = \dim\{\ker(\underline{D}^*)\}$, but we shall express it in terms of a normalized cycle basis [4] defined below and denote it as the *cross differential* of the graph. The operators $\underline{D}, \underline{D}^*, \underline{D}_\times$ then bear an obvious resemblance to grad, div and rot (or curl) of vector calculus, a resemblance made more compelling below.

With power given by $\dot{W} = (\underline{f}, \underline{u})$, we can formulate a virtual work principle in terms of the above operators as follows (cf. [32]). Designating column χ as *compatible* if it can be written as a nodal difference $\chi = \underline{D}\varphi$ and as a *conserved flow* if it satisfies $\underline{D}^*\chi = \underline{0}$, we obtain the associated conservation-compatibility duality [6, 32]

$$\dot{W} = (\underline{f}, \underline{D}\,\varphi) = (\underline{D}^*\underline{f}, \varphi) = 0 \ \forall\varphi, \text{ iff } \underline{D}^*\underline{f} = \underline{0} \tag{15}$$

and

$$\dot{W} = (\underline{D}_\times\psi, \underline{u}) = (\psi, \underline{D}^*_\times\underline{u}) = 0 \ \forall\psi, \text{ iff } \underline{D}^*_\times\underline{u} = \underline{0}, \tag{16}$$

i.e., compatibility implies conservation and vice versa.

Similar relations apply to arrays of space vector and tensors $\mathbf{x} = [\mathbf{x}^i]^*$ and $\underline{\mathbf{A}} = [\mathbf{A}^{ij}]$, with scalar product and adjoint

$$(\underline{\mathbf{A}}\,\mathbf{y}, \mathbf{z}) := (\underline{\mathbf{A}}\,\mathbf{y})^* \cdot \mathbf{z} = (\mathbf{y}, \underline{\mathbf{A}}^*\mathbf{z}), \text{ with } \mathbf{A}^{*ij} = \mathbf{A}^{Tji}. \tag{17}$$

Thus, for the Satake graph $\mathcal{G}$, the substitutions $\underline{f} \to \underline{\mathbf{f}}$, and $\underline{u} \to \underline{\mathbf{u}}$ in (15)–(16) yields the quasi-static equilibrium of forces and the compatibility of relative velocities (or displacements) [31, 38]

$$\underline{D}^*\underline{\mathbf{f}} = \underline{\mathbf{0}}, \text{ and } \underline{D}^*_\times\underline{\mathbf{u}} = \underline{\mathbf{0}}. \tag{18}$$

The second relation of course is satisfied identically by the substitution $\varphi \to \mathbf{v}$ in (15)–(16), where $\mathbf{v} = \dot{\mathbf{x}}$ denotes (nodal) velocities (or infinitesimal displacements) of grain centroids $\mathbf{x} = [\mathbf{x}^i]^*$, connected by branch or edge vectors $\mathbf{l} = [\mathbf{l}^i]^* = \underline{D}\mathbf{x}$. The first is satisfied by the substitution in (15)–(16) of a vector array for a scalar array $\psi \to \boldsymbol{\psi}$, providing a discrete analog of the Helmholtz representation of solenoidal vector fields.

The rank of $\underline{D}^*$ equals $N - 1$ the number of independent scalar balances in the last member of (15) or vector balances in the first member of (18), and the rank of $\underline{D}^*_\times$ equals $M = E - N + 1$, which follows from the celebrated Descartes–Euler polyhedral formula and also from the later analysis of Kirchhoff [7, 29] for electrical networks. The latter contains the notion of a cycle basis for conserved flows, which provides a particularly attractive null-space basis for $\underline{D}^*$, namely the normalized *irreducible cycle basis* or *mesh* for

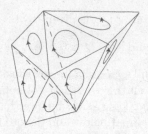

Fig. 2. Meshes for a 3d simplicial complex

the associated graph $\mathcal{G}$. This consists of the maximal linearly independent set of cyclic currents $\underline{f} = [f_1, \dots, f_M]^*$ having as their only non-zero components f_k, such that $|f_k| = 1$ on branches k forming irreducible (or "elementary" [4]) cycles or "meshes", i.e. cycles that contain no other cycles, on $\mathcal{G}$, as illustrated in Fig. 2 for the polyhedral graph defined by a 3d simplicial complex (i.e. face-connected cluster).

We then take $\underline{D}_\times$ to be the $E \times M$ matrix (transpose of that denoted by L in [38])

$$D_\times^{ij} = \begin{cases} +1, \text{if edge } i \text{ is coincident and confluent with cycle } j, \\ -1, \text{if edge } i \text{ is coincident and not confluent with cycle } j, \\ 0, \text{otherwise.} \end{cases} \quad (19)$$

This imparts the status of difference operator and allows for a symmetric duality in the case of planar graphs, as discussed next.

A Note on Duality

The notion of duality is ubiquitous and varied in the literature on graphs and geometry [4, 12], and the following paragraphs represent an attempt to distinguish some special cases particularly relevant to the subject at hand.

In the case of the planar graph $\mathcal{G}$, e.g. associated with planar electrical networks or idealized 2d granular media [31], one can identify a dual graph $\mathcal{G}'$ with node–mesh (vertex–face) duality defined by

$$M' = N - 1, \; N' = M + 1, \; E' = E \quad (20)$$

and illustrated by enumerated edges (e), nodes (n) and meshes (m) for the portion of a graph shown in Fig. 3. (The graph on the right is obtained from that on the left by letting the nodes n expand, the meshes m shrink, and the edges e rotate, and vice versa.)

Figure 4 illustrates the further grad–rot, force–flow and potential–stream function duality for compatible forces $\underline{u}$ and conserved flows $\underline{f}$

$$\underline{D}' = \underline{D}_\times, \; \underline{D}_\times' = \underline{D}, \; \underline{u}' = \underline{f}, \; \underline{f}' = \underline{u}, \; \underline{\varphi}' = \underline{\psi}, \; \underline{\psi}' = \underline{\varphi}, \quad (21)$$

with

$$\underline{u} = \underline{D}\,\underline{\varphi}, \text{ and } \underline{f} = \underline{D}_\times\,\underline{\psi} \quad (22)$$

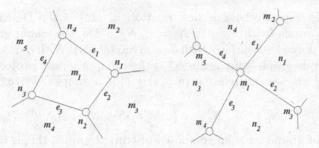

Fig. 3. Node–mesh duality for a planar graph

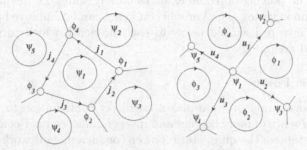

Fig. 4. Potential–function duality

This highly symmetric duality does not carry over to non-planar graphs such as those associated with 3d polyhedra and simplicial complexes, because edges generally are contiguous with more than two faces (or cycles). A duality that preserves edges corresponds to a hypergraph structure [4], with "hyper-edges" connecting more than two nodes. However, another form of duality is possible.

According to a (Schäfli–Poincaré) formula for d-dimensional polytopes (generalized polyhedra) [12, 48], we have

$$\sum_{m=0}^{d-1}(-1)^m N_m = I_d, \tag{23}$$

where N_k denotes the number of constituent elements of dimension k ($k = 0$ for vertices or nodes, $k = 1$ for "edges" connecting vertices, $k \in [2, d]$ for hyperfaces), and I_d is a topological invariant depending on the connectivity of the underlying manifold.[6] For simply connected manifolds $I_d = 1 - (-1)^d$, giving

$$N_0 - N_1 + N_2 = 2, \ \ N_0 - N_1 + N_2 - N_3 = 0, \tag{24}$$

---

[6] The actual value of I_d is less important for the present purposes than the fact that $\Delta I_d = 0$ for any addition or deletion of nodes, edges, etc., that preserves connectivity in an existing simplicial complex.

for $d = 3, 4$, respectively. The first relation in (24) is the Descartes–Euler polyhedral formula (with $N_0 = N, N_1 = E, N_2 = M$), the planar graph representing a 3d polyhedral surface, with dual given by (20). With our 3d simplicial complex and its graph being regarded as the an appropriate projection of the 4d polytope, the second relation in (24) yields a simplex–vertex/edge–face duality

$$N_0' = N_3, \ N_1' = N_2, \ N_2' = N_1, \ N_3' = N_0, \tag{25}$$

In the case of granular media composed of convex grains, this is tantamount to the oft-invoked duality [1] between the Voronoi polyhedron centered on a grain and the polyhedral complex of Delaunay simplices having the same center as common vertex. The Voronoi construct can be employed [1] to assign the vectorial area **a** discussed below to each branch **l**, which corresponds to the subtended area in [2].

2.3 Extrinsic Power

As shown above, grain rotation makes no direct contribution to quasi-static work in the absence of body couples and intergranular contact couples. Under these circumstances, the quasi-static power (or incremental work) of contact forces for a granular assembly is given by

$$\dot{W} = (\underline{\mathbf{f}}, \underline{\mathbf{u}}) := \underline{\mathbf{f}}^* \cdot \underline{\mathbf{u}} = \sum_{k=1}^{E} \mathbf{f}^k \cdot \mathbf{u}^k. \tag{26}$$

The condition $\dot{W} = 0$, analogous to Tellegen's theorem [40] for electrical circuits, requires external forcing of some subset of particles.[7] Otherwise, it represents the virtual-work principle of (15)–(16).

To pass from contact network (weighted graph) to continuous manifold, we provisionally associate branches $\underline{\mathbf{l}} = \underline{D}\,\mathbf{x}$ with the tangent space, and vectors $\underline{\mathbf{d}} = [\mathbf{d}_k]^*$, $\mathbf{d}^k = \mathbf{a}^k/V^k$, with the cotangent space. As explained below in the Appendix, $\mathbf{a}^k$ denotes a vectorial area and $V^k = \mathbf{l}^k \cdot \mathbf{a}^k$ a volume associated with *simplicial edge complex* (cluster of contiguous Delaunay simplices with common edge $\mathbf{l}^k$). Thus,

$$\underline{\mathbf{x}} \Rightarrow \mathbf{x}, \ \underline{\mathbf{l}} = \underline{D}\,\underline{\mathbf{x}} \Rightarrow d\mathbf{x}, \tag{27}$$

hence

$$\underline{D}\,\underline{\varphi} = (\underline{D}\,\underline{\mathbf{x}}, \underline{D}\,\underline{\varphi}) \Rightarrow d\varphi(\mathbf{x}) = d\mathbf{x} \cdot \nabla\varphi, \tag{28}$$

where

$$\underline{\mathbf{D}} := [\mathbf{d}^i D^{ij}] \Rightarrow \nabla \tag{29}$$

---

[7] The substitution $\dot{W}, \mathbf{v}, \mathbf{u}, \mathbf{f} \to V, \mathbf{x}, \mathbf{l}, \mathbf{a}$ in (26), **a** denoting a vectorial area discussed below, yields a geometric formula relevant to granular compaction and dilatancy. By choosing **l**, **a** as primary variables in the maximum-entropy estimates discussed in [20], one obtains formulae similar to those proposed in [8, 28].

The latter is a special case of a higher-order gradient

$$\underline{\mathbf{D}}^{(n)} := [\mathbf{d}^{in}D^{ij}] \Rightarrow \boldsymbol{\nabla}^n, \text{ with } \mathbf{d}^{in} := (\mathbf{d}^i)^n, \; n = 1, 2, \cdots , \qquad (30)$$

subject to improvement through the replacement of D^{ij} by a more general (finite-difference) approximation $D^{(n)ij}$, say, based on a connected set of branches. As it stands, (30) is adequate for a continuum interpretation of the results to follow.

3 Energy-Based Homogenization

In the following, we let $\langle \chi \rangle_c, \langle \chi \rangle_\phi$ denote, respectively, the number averages of the components χ^k of array $\underline{\chi} = [\chi^k]^*$ over branches or edges $k \in [1, E]$ and volume averages over the associated edge complexes, such that

$$\langle \chi \rangle_c := \frac{1}{E} \sum_k \chi^k, \; \langle \chi \rangle_\phi := \frac{1}{V} \sum_k V^k \chi^k, \text{ and } \langle \chi \rangle_\phi = n_c \langle V\chi \rangle_c, \qquad (31)$$

where $n_c = E/V$ denotes branch (or total contact) density and $\underline{V\chi} = [V^k \chi^k]^*$.

Then, $\dot{W}$ in (26) may be assumed to arise from boundary forces that provide the stress power

$$\dot{w} = \frac{\dot{W}}{V} = \langle \mathbf{T} : \mathbf{L} \rangle_\phi = \frac{1}{V} \sum_{k=1}^{E} V^k \mathbf{T}^k : \mathbf{L}^k, \qquad (32)$$

the analog of (3), where the tensor products [1, 20]

$$\mathbf{T}^k = \frac{1}{V^k} \mathbf{f}^k \otimes \mathbf{l}^k, \; \mathbf{L}^k = \frac{1}{V^k} \mathbf{u}^k \otimes \mathbf{a}^k, \text{ with } V^k = \mathbf{l}^k \cdot \mathbf{a}^k, \qquad (33)$$

represent contributions of branches to the global averages $\langle \mathbf{T}_c \rangle, \langle \mathbf{L} \rangle_c$ and, hence, to $\langle \mathbf{T}_\phi \rangle, \langle \mathbf{L} \rangle_\phi$ As is the case with other heterogeneous media. $\dot{w}$ in (32) generally is not given by $\langle \mathbf{T} \rangle_\phi : \langle \mathbf{L} \rangle_\phi$, owing to macroscopic gradients and random microscopic fluctuations.

In keeping with the continuum form (1), and following previous works, we assume that $\mathbf{f}^k, \mathbf{u}^k$ are known, e.g. given by a micromechanical theory or a numerical simulation, and we fit the known data for $\mathbf{u}_k$ with

$$\mathbf{u}^k = \tilde{\mathbf{u}}^k + \mathbf{u}^{k\prime}, \text{ with } \tilde{\mathbf{u}}^k = \tilde{\mathbf{L}}\mathbf{l}^k + \tilde{\mathbf{L}}_2 : \mathbf{l}^{k2} + \ldots + \tilde{\mathbf{L}}_m : \mathbf{l}^{km}, \qquad (34)$$

where $\mathbf{l}^{kj} = (\mathbf{l}^k)^j$. The polynomial in $\mathbf{l}^k$ is attributed to macroscopic gradients and $\mathbf{u}^{k\prime}$ to random fluctuations.

To specify the parameters $\tilde{\mathbf{L}}_n$ in (34) for some subset of m branch vectors $\mathbf{l}^k$, several authors advocate strict polynomial fits, with a maximal value of m, or else some other "best" fit of (34) to continuum kinematics [27, 33, 42]. Given the well-known pathology ("overfitting") of polynomial fits of random data, and in view of the paramount importance of energy, we take the position that a "best" fit should rather be based on minimization of an appropriate norm of stress-power fluctuations plus some norm of the variation implied by (34), in the spirit of the so-called "generalized additive models" (GAM) [49].

As a prototypical linear method, consider

$$\sigma^2 = (\underline{\mathbf{u}}', \underline{\underline{\mathbf{G}}}\underline{\mathbf{u}}') + Q, \quad \partial\sigma^2/\partial\tilde{\mathbf{L}}_n = 0, \ n = 1, 2, \ldots, m, \tag{35}$$

where the Q denotes a quadratic form in the $\tilde{\mathbf{L}}_n$. This leads to a set of linear equations for $\tilde{\mathbf{L}}_n$, with corresponding estimate for stress power

$$\tilde{w} = \tilde{\mathbf{T}}{:}\tilde{\mathbf{L}} + \tilde{\mathbf{T}}_2{:}\tilde{\mathbf{L}}_2 + \ldots + \tilde{\mathbf{T}}_m{:}\tilde{\mathbf{L}}_m, \tag{36}$$

where, as the analog of (4) and in a form proposed elsewhere [19], the moment stresses are given by the average moments (multipoles)

$$\tilde{\mathbf{T}}_n = n_c\langle \mathbf{f} \otimes \mathbf{l}^n \rangle_c = \frac{1}{V} \sum_{k=1}^{m} \mathbf{f}^k \otimes \mathbf{l}^{kn}, \ n \in [1, m], \tag{37}$$

irrespective of the resulting solution for $\tilde{\mathbf{L}}_n$.

Although not explored in detail here, one plausible form for $\underline{\underline{\mathbf{G}}}$ in (35) is

$$\underline{\underline{\mathbf{G}}} = \text{diag}[\mathbf{1} - \hat{\mathbf{f}}^k \otimes \hat{\mathbf{f}}^k], \text{ where } \hat{\mathbf{f}} = \mathbf{f}/|\mathbf{f}| \ . \tag{38}$$

This penalizes fluctuations $\underline{\mathbf{u}}'$ that do no work, representing a loose analogy to thermal fluctuations in molecular systems. One obtains a dual for (34)–(37) by taking Q to be a quadratic form in $\tilde{\mathbf{T}}_n$, followed by the interchanges

$$\mathbf{u}^k \leftrightarrow \mathbf{f}^k/V^k, \ n!\tilde{\mathbf{L}}_n \leftrightarrow \tilde{\mathbf{T}}_n, \ \mathbf{l}^k \leftrightarrow \mathbf{d}^k := \mathbf{a}^k/V^k, \tag{39}$$

where $\mathbf{d}^{kn}{:}\mathbf{l}^{kn} = 1$, so that

$$\tilde{\mathbf{f}}^k = V^k(\tilde{\mathbf{T}}\mathbf{d}^k + \tilde{\mathbf{T}}_2{:}\mathbf{d}^{k2} + \ldots + \tilde{\mathbf{T}}_m{:}\mathbf{d}^{km}), \tag{40}$$

which corresponds to the so-called "static hypothesis" for contact forces of [10]. The resulting (dual) estimate for $\tilde{\mathbf{L}}_n$ is

$$\tilde{\mathbf{L}}_n = \langle \mathbf{L}_n \rangle = \frac{1}{n!} \sum_{k=1}^{m} \mathbf{u}^k \otimes \mathbf{d}^{kn}, \ n \in [1, m] \tag{41}$$

with

$$\underline{\mathbf{L}}_n = \frac{1}{n!}[\mathbf{u}^k \otimes \mathbf{d}^{kn}] = \frac{1}{n!}\underline{\mathbf{D}}^{(n)} \otimes \underline{\mathbf{v}}, \tag{42}$$

where $\underline{\mathbf{D}}^{(n)}$ is the matrix defined in (30). The first term in (41) is equivalent to the volume averages couched elsewhere [1, 2] in terms of infinitesimal displacement gradients.

Note that, depending on the form of Q, the estimates obtained for the moment stresses $\tilde{\mathbf{T}}_n$ in (40) would not necessarily agree with those of (37). Similarly, the velocity gradients in (35) are not necessarily the same as those given in (41).

More generally, on replacing Q in (35) by a quadratic form in both $\tilde{\mathbf{L}}^k, \tilde{\mathbf{T}}^k$, and employing similar polynomial representations of $\tilde{\mathbf{u}}^k, \tilde{\mathbf{f}}^k$ in terms of $\mathbf{l}^k, \mathbf{d}^k$, respectively, we obtain a more general, simultaneous (GAM) estimate of stresses and gradients. Bilinearity of Q in $\tilde{\mathbf{L}}^k, \tilde{\mathbf{T}}^k$ might allow for an interpretation in terms of energy.[8]

Again, the results for $\tilde{\mathbf{T}}_n, \tilde{\mathbf{L}}_n$ would not necessarily agree with those obtained above, and it should be amply clear that the definitions of moment stresses and kinematic gradients depend both on the nature of the objective function Q and the inhomogeneity in contact forces and branch vectors.

3.1 Intrinsic Moments and Continuum Fields

The preceding discussion deals with extrinsic quantities defined by the motion of grain centroids under the action of intergranular contact forces. The treatment of localized contact mechanics (as in the Hertzian elastic contact), as well as the treatment of intrinsic quantities such as global particle deformation would require a consideration of the internal mechanics of individual grains, which one usually assumes to consist of a simple continuum endowed with appropriate constitutive equations, elastic, viscoelastic, elastoplastic, etc.

The detailed treatment of micromechanics is beyond the scope of the present article, which is rather concerned with general aspects and consequences. We merely note that the effective particle stress $\mathbf{T}^p$ for a particle p is given by (7), Higher moment stresses are given by a reinterpretation of (37) in which branch vectors $\mathbf{l}$ are replaced by moment arms $\mathbf{r}$. In a similar way, (35) gives a similar but less exact estimate of velocity gradients, by interpreting $\mathbf{u}$ as relative velocity between contact point k and particle centroid, and by basing $\mathbf{d}^k$ on an effective contact area. The latter description of particle kinematics represents a type of finite-element approximation, whereas an exact treatment of the micromechanics would generally involve solving field equations for the particle interior, subject to localized tractions on the particle surface, followed by appropriate averaging of solutions over particle volume.

At any rate, it is clear that the localized surface stresses provide a coupling of the intrinsic modes to the extrinsic modes represented by motion of particle centroids. This paramount aspect of granular mechanics may be obscured by the usual micromechanical analysis, where particle rotation, a property of finite grains, is placed ab initio on the same footing as the motion of particle centroids.

---

[8] The preceding paragraph corrects several errors in the corresponding paragraph of [21].

With an appropriate replacement of (34) and (35)–(41), one obtains higher-order micromorphic effects, represented by $\mathbf{T}_n^p$, $\mathbf{L}_n^p$, $n > 1$. The ever-increasing dependence on particle length scales is thereby manifest. In a similar vein, we expect that higher-order contact moments will exhibit a similar dependence on the dimension of contact zones.

Given the above estimates of continuum-level moments, the following formula is suggested by, but not rigorously derived from the above-cited works of Eringen and coworkers:

$$\mathbf{X} = \nu^c \mathbf{X}^c + \nu^p \mathbf{X}^p + \mathbf{X}^s, \tag{43}$$

where $\mathbf{X} = \langle \mathbf{T}_n \rangle$ or $\langle \mathbf{L}_n \rangle$, $n = 1, 2 \ldots$, represent volume (or surface) averages, with $\langle \mathbf{X}_n \rangle^i = O(1)$ for $\nu^i \to 0$. The superscript c refers to a continuum-level contribution arising from the relative motion of particle centroids; p to a contribution arising from the internal structure of particles, regarded as pieces of a continuous medium; and s to a contribution arising from singular surfaces. ν^p denotes particle volume fraction and ν^c void fraction, given by $1 - \nu^p$ in the usual granular medium.

Typical singular surfaces involve interfacial slip, such as cracks, or other kinematic discontinuities, or interfacial tension and other (multipolar) stress jumps [16]. The relation (43) appears to cover various limiting case, e.g. $\nu^p \to 0, 1$, and Mindlin's special case $\mathbf{X}^c \to \mathbf{X}^p$ [35], often used for multipolar elasticity.

Although the moment stress power is generally not given by $\langle \mathbf{T}_n^i \rangle^i : \langle \mathbf{L}_n \rangle^i$, one may readily obtain the following generalization of a well-known result for $m = 1$ from (5): For a graded material of grade m, with $\mathbf{L}_k^i = (\nabla^k \mathbf{w})^T / k!$ for $k \in [1, m]$, the stress power $\dot{w}_m$ is given by

1. $\langle \mathbf{T}_m \rangle : \overline{\mathbf{L}}_m$ for velocity fields $\mathbf{w}$ which have boundary values given by a polynomial of maximal degree m, with $\overline{\mathbf{L}}_m^i = (\nabla^m \mathbf{w})^T / m!$, a constant over ∂V, and by
2. $\overline{\mathbf{T}}_m : \langle \mathbf{L}_m \rangle$ for moment-stress fields whose surface (moment) tractions satisfy $\mathbf{T}_k \mathbf{n} = \delta_{km} \overline{\mathbf{T}}_k \mathbf{n}$, with $\overline{\mathbf{T}}_m$ constant over ∂V.

Since this result applies to any finite cluster of particles, subject to inhomogeneous conditions of displacement or effective stress on the periphery, one concludes that volume averages do not provide a proper definition of continuum fields in highly inhomogeneous assemblies. This casts considerable doubt on the use of (7) and, hence, of (9) to define Cauchy stress, the latter of which goes back, in the field of granular mechanics at least to Weber [3, 47] and, in theoretical elasticity, to Cauchy [9] (cf. note B in the appendix of [34], discussed in [18] and also in [26]). The breakdown of (7) is suggested by the mean-field theory of Jenkins [26] for elastic-sphere assemblies, the theory of Bardet and Vardoulakis [3] for small granular assemblies, and also by more recent calculations [13]. It is worth noting that the theory of [26] involves gradients in contact force reminiscent of (40) above, whereas the other results [3, 13] depend on boundary effects in small samples.

4 Conclusions

A synthesis has been presented of graph-theoretic methods and energy-based homogenization to derive continuum models of discrete granular media. As anticipated by several previous workers, it is concluded that the multipolar continuum, either graded or micromorphic, represents a plausible model for the typical granular medium.

It has been shown that the special case of a graded continuum, including the simple (grade one) continuum, is defined solely by the extrinsic modes associated with the motion of grain centroids, in contrast to the micrormorphic continuum, which arises from intrinsic modes represented by the internal mechanics of grains.

Within the subclass of micromorphic continua, the micropolar (Cosserat) limit is appropriate for nearly rigid grains. However, In the absence of intergranular contact moments, it has been demonstrated above that grain rotation makes no direct contribution to quasi-static contact work, and that the widely accepted formula based on volume averaging yields a symmetric Cauchy stress. One therefore concludes that the emergence of Cosserat effects implies the breakdown of this formula. Otherwise, the existence of moment stress must be attribute to kinematic gradients, suggesting that the graded continuum may prove to be more appropriate than the micropolar continuum for the quasi-static mechanics of rigid granular media.

There remain open questions as to the validity of the multipolar balances based on [19], the interpretation of Eringen's micromorphic theory in terms of volume averages, and the extension to granular dynamics. Incidentally, given the previous works [16] on micromorphic continua, the latter appears quite feasible.

As pointed out previously [21], further investigations of shear bands and of short-wavelength shear waves should provide a plausible testing ground for multipolar theories.

Appendix: Simplex and Edge-Complex Gradients

The following, an elaboration on the method employed in [19], serves to define gradients and various geometrical properties associated with simplicial complexes. In a (Euclidean) space of dimension d, we define the simplicial gradient of a function $\underline{\varphi} = [\varphi^k]^*$, specified on the $d + 1$ vertices $\mathbf{x}^k$ of a simplex s, by means a linear interpolation $\varphi(\mathbf{x})$ based on barycentric coordinates $\xi^k(\mathbf{x})$ [12,32] (affine functions of $\mathbf{x}$ defined below[9]), with

$$\varphi(\mathbf{x}) = \sum_{k=1}^{d+1} \xi^k \varphi^k, \ \sum_{k=1}^{d+1} \xi^k = 1, \ \xi^k \in [0,1], \ \xi^k(\mathbf{x}^k) = 1, \quad (44)$$

$$\nabla\varphi = \sum_{k=1}^{d+1} \mathbf{g}^k \varphi^k, \ \mathbf{g}^k := \nabla\xi^k \text{ (const.)}, \ \sum_{k=1}^{d+1} \mathbf{g}^k = \mathbf{0}, \quad (45)$$

so that

$$\nabla\varphi = \sum_{k=1}^{d+1} \mathbf{g}^k(\varphi^k - \varphi^o), \ o \in [1, d+1]. \quad (46)$$

The last member of (45) is merely Green's theorem for simplex s, since

$$\mathbf{0} = \int_{V^s} \nabla(1)dV = \int_{\partial V^s} (1)ds \equiv \sum_{k=1}^{d+1} \mathbf{s}^k, \text{ with } \mathbf{s}^k = 2V^s\mathbf{g}^k. \quad (47)$$

The vector $\mathbf{s}^k$ is normal to facet (i.e. a bounding hyperplane of dimension $d-1$) k, having magnitude $|\mathbf{s}^k|$ equal to its $(d-1)$-volume, and V^s is the d-volume of the simplex. (The formulae presented here follow from a consideration of the linear map that carries a standard d-simplex, i.e. one-half the unit d-hypercube, into an arbitrary d-simplex.)

Given a basis composed of d edge vectors $\mathbf{g}_k = \mathbf{l}^k$ [19], an appropriate set d of the $d+1$ vectors $\mathbf{g}^k$ provide a reciprocal basis, with $\mathbf{g}^i \cdot \mathbf{g}_j = \delta^i_j$. This is illustrated by the special case $\varphi \equiv \mathbf{x}$ in (44), yielding by (46) a well-known expression for the unit tensor

$$\mathbf{1} = \sum_{k=1}^{d+1} \mathbf{g}^k \otimes \mathbf{g}_k, \text{ with } \mathbf{g}_k = \mathbf{x}^k - \mathbf{x}^o, \text{ for } o, k \in [1, d+1]. \quad (48)$$

In this representation, $\mathbf{x}^o$ serves as origin for $\mathbf{g}_k$, $k \in [1, d+1]\backslash o$, with $\mathbf{g}_j$, $j \neq k$, lying in the facet normal to $\mathbf{g}^k$. This is illustrated for $d = 3, o = 4$ in Fig. 5.

The further special case $\varphi \equiv \xi^k$ in (44) yields an explicit formula for barycentric coordinates

$$\xi^k(\mathbf{x}) = \mathbf{g}^k \cdot (\mathbf{x} - \mathbf{x}^k) + 1, \text{ for } k \in [1, d+1], \quad (49)$$

with $\xi^k(\mathbf{x}^j) = \mathbf{g}^k \cdot (\mathbf{x}^j - \mathbf{x}^k) + 1 = \mathbf{g}^k \cdot (\mathbf{g}_j - \mathbf{g}_k) + 1 = \delta^k_j$.

With index s enumerating simplices, (46) becomes

$$\nabla\varphi^s = \sum_k \mathbf{g}^{sk}(\varphi^{sk} - \varphi^{so}) \equiv \frac{1}{V^s} \sum_k \mathbf{s}^{sk}(\varphi^{sk} - \varphi^{so}),$$

---

[9] $\xi^k = \cos^2\theta_k$, $\theta_k \in [0, \pi/2]$, provide part of a branched covering [39] of the $(d+1)$-sphere surface by a d-simplex.

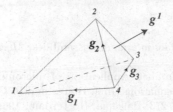

Fig. 5. Elements of an edge basis and its reciprocal facet basis for a 3d simplex

yielding for the volume average gradient over an assembly

$$\langle \nabla \varphi \rangle = \frac{1}{V} \sum_s V^s \nabla \varphi^s = \frac{1}{V} \sum_s \sum_k s^{sk}(\varphi^{sk} - \varphi^{so}), \tag{50}$$

where ranges on summations are understood.

For clarity we restrict the discussion to $d \leq 3$ and define a *simplicial edge complex* (or edge cluster) $\sigma(e)$ to be the set of simplices having common edge e, with $c = \{ko\}$ in (50). Then, on rearranging summations and recalling the definition of the matrix differential $\underline{D} = [D^{en}]$, we may express (50) as

$$\langle \nabla \varphi \rangle = \frac{1}{V} \sum_{e=1}^{E} V^e \langle \nabla \varphi \rangle^e, \quad \text{where } \langle \nabla \varphi \rangle^e = \sum_{n=1}^{N} \mathbf{d}^e D^{en} \varphi^n, \tag{51}$$

with

$$\mathbf{d}^e = \mathbf{a}^e/V^e, \quad \mathbf{a}^e = \sum_{s \in \sigma(e)} \mathbf{s}^s \equiv 2 \sum_{s \in \sigma(e)} V^s \mathbf{g}^e. \tag{52}$$

The second relation in (51) obviously can be written as $[\langle \nabla \varphi \rangle^e]^* = \underline{D}\underline{\varphi}$, which establishes the relation (29).

The volume V^e introduced here is arbitrary and could be chosen as the sum of simplex volumes V^s, $s \in \sigma(e)$. However, for purposes of defining volume averages, it seems more appropriate to employ disjoint volumes, either by reference to particle-based Voronoi cells [1,2] or related particle-free geometric constructs.

Acknowledgement

I am grateful to several colleagues, notably J-J. Moreau, F. Radjai, and N. Rivier, for various discussions of ideas and for bibliographic references. Gratitude is also due to the organizers of the 2005 workshop on granular materials in the Centre Borel, Institut Henri Poincaré, Paris, for providing a tranquil but stimulating environment for some of the effort reflected in this work.

References

1. K. Bagi. Stress and strain in granular assemblies. *Mech. Mater.*, 22(3):165–177, 1996.
2. J. P. Bardet and J. Proubet. Application of micromechanics to incrementally nonlinear constitutive equations. In J. Biarez and R. Gourvès, editors, *Powders and Grains*, pages 265–273. Balkema, Rotterdam, 1989.
3. J. P. Bardet and I. Vardoulakis. The asymmetry of stress in granular media. *Int. J. Solids Struct.*, 38(2):353–67, 2001.
4. C. Berge. *Graphs and Hypergraphs*. North-Holland Publishing Co., Amsterdam, 1973.
5. N. Biggs. *Interaction Models*. Cambridge University Press, Cambridge, U.K., 1977.
6. N. Biggs. *Algebraic Graph Theory*. Cambridge University Press, Cambridge, U.K., 2nd edition, 1994.
7. N. Biggs, E. K. Lloyd, and R. J. Wilson. *Graph Theory* 1736–1936. Clarendon Press, Oxford, U.K., 1976.
8. R. Blumenfeld and S. F. Edwards. Granular entropy: Explicit calculations for planar assemblies. *Phys. Rev. Lett.*, 90(11):114303/1–4, 2003.
9. A. L. Cauchy. De la pression ou tension dans un système de points matériels. *Exércises de maths.*, 2:42, 1827.
10. C. S. Chang and J. Gao. Kinematic and static hypotheses for constitutive modelling of granulates considering particle rotation. *Acta Mech.*, 115(1–4):213–229, 1996.
11. E. Cosserat and F. Cosserat. *Théorie des Corps Déformables* (Theory of Deformable Bodies). A. Hermann et Fils, Paris, 1909. (English translation in NASA TT F-475 11,561, Nat. Aero. Space Adm., Washington D.C., February, 1968).
12. H. S. M. Coxeter. *Regular Polytopes*. Dover, New York, 3rd edition, 1973.
13. W. Ehlers, E. Ramm, S. Diebels, and G. A. D'Addetta. From particle ensembles to Cosserat continua: homogenization of contact forces towards stresses and couple stresses. *Int. J. Solids Struct.*, 40(24):6681–6702, 2003.
14. A. C. Eringen. Theory of micropolar elasticity. In *Fracture, an advanced treatise*, volume 2, pages 622–728. Academic Press, New York, 1968.
15. A. C. Eringen. Balance laws of micromorphic continua revisited. *Int. J. Eng. Sci.*, 30(6):805–10, 1992.
16. A. C. Eringen. *Microcontinuum Field Theories: Foundations and Solids*. Springer-Verlag, New York-Berlin, 1999.
17. A. C. Eringen. *Nonlocal Continuum Field Theories*. Springer, New York, 2002.
18. J. D. Goddard. Microstructural origins of continuum stress fields - a brief history and some unresolved issues. In D. De Kee and P. N. Kaloni, editors, *Recent developments in sturctured continua*, volume 143 of *Pitman Research Notes in Mathematics*, pages 179–208. Longman/J. Wiley, New York, 1986.
19. J. D. Goddard. Continuum modeling of granular assemblies. In H. J. Herrmann, et al., editor, *NATO ASI, Physics of Dry Granular Media*, page 24. Kluwer, Dordrecht, 1998.
20. J. D. Goddard. On entropy estimates of contact forces in static granular assemblies. *Int. J. Solids Struct.*, 41(21):5851–61, 2004.

21. J. D. Goddard. Granular media as generalized micromorphic continua. In R. García-Rojo and et al., editors, *Powders and Grains 2005*, volume 1, pages 129–134. Taylor & Francis Group, London, 2005.
22. A. E. Green and P. M. Naghdi. A unified procedure for construction of theories of deformable media. 1.classical continuum physics & 2.generalized continua. *Proc. Roy. Soc. London A*, 448(1934):335–356, 357–377, 1995.
23. A. E. Green and R. S. Rivlin. Multipolar continuum mechanics. *Arch. Rat. Mech. Anal.*, 17(2):113–47, 1964.
24. A. E. Green and S. Rivlin. Relation between director and multipolar theories in continuum mechanics. *ZAMP*, 18(2):208–18, 1967.
25. A. E. Green and R. S. Rivlin. Simple force and stress multipoles. *Arch. Rat. Mech. Anal.*, 16:325–53, 1964.
26. J. T. Jenkins. Anisotropic elasticity for random arrays of identical spheres. In J. Wu, T. C. T. Ting, and D. M. Barnett, editors, *Modern Theory of Anisotropic Elasticity and Applications*, pages 368–377. SIAM, Philadelphia, 1990.
27. J. T. Jenkins and M. A. Koenders. The incremental response of random aggregates of identical round particles. *Eur. Phys. J. E*, 13(2):113–23, 2004.
28. K-I. Kanatani. An entropy model for shear deformation of granular materials. *Lett. Appl. Engng. Sci. (Int. J. Eng. Sci.)*, 18:989–998, 1980.
29. G. Kirchhoff. Üeber die Auflösung der Gleichungen, auf welche man bei der Untersuchung der linearen Vertheilung galvanischer Ströme gefürt wird. *Ann. Phys. Chem.*, 72(12):497 508, 1847. (English translation in Biggs 1976).
30. N. Kirchner and P. Steinmann. A unifying treatise on variational principles for gradient and micro-morphic continua. *Phil. Mag.*, 85:3875–95, 2005.
31. N. P. Kruyt. Statics and kinematics of discrete Cosserat-type granular materials. *Int. J. Solids Struct.*, 40(3):511–534, 2003.
32. S. Lefschetz. Applications of Algebraic Topology - Graphs and Networks: The Picard-Lefschetz Theory and Feynman Integrals. In *Applied mathematical sciences (Springer-Verlag New York Inc.)*, volume 16. Springer-Verlag, New York, 1975.
33. C.-L. Liao, T.-P. Chang, D.-H. Young, and C.·S. Chang. Stress-strain relationship for granular materials based on the hypothesis of best fit. *Int. J. Solids Struct.*, 34(31–32):4087–100, 1997.
34. A. E. H. Love. *A Treatise on the Mathematical Theory of Elasticity*. Dover, New York, 4th edition, 1944. (Bibliographic endnotes).
35. R. D. Mindlin. Micro-structure in linear elasticity. *Arch. Rat. Mech. Anal.*, 16(1):51–78, 1964.
36. H. B. Mühlhaus and I. Vardoulakis. Thickness of shear bands in granular materials. *Geotechnique*, 37(3):271–283, 1987.
37. M. Oda and K. Iwashita, editors. *Mechanics of Granular Materials - An Introduction*. Balkema, Rotterdam/Brookfield, 1999.
38. M. Satake. New formulation of graph-theoretical approach in the mechanics of granular-materials. *Mech. Mater.*, 16(1–2):65–72, 1993.
39. I. V. Savel'ev. Branched coverings over manifolds. *J. Math. Sci.*, 119:605–57, 2004.
40. O. Shai. Deriving structural theorems and methods using Tellegen's theorem and combinatorial representations. *Int. J. Solids Struct.*, 38(44–45 Oct 12):8037–52, 2001.
41. E. S. Suhubi and A. C. Eringen. Nonlinear theory of micro-elastic solids. II. *Int. J. Eng. Sci.*, 2(4):389–404, 1964.

42. A. S. J. Suiker, R. De Borst, and C. S. Chang. Micro-mechanical modelling of granular material. part 1: Derivation of a second-gradient micro-polar constitutive theory. *Acta Mech.*, 149(1–4):161–180, 2001.

43. C. Truesdell and W. Noll. The non-linear field theories of mechanics. In S. Flügge, editor, *Encyclopedia of Physics (Handbuch der Physik)*, volume III/3. Springer-Verlag, Berlin, New York, 1965.

44. C. Truesdell and R. A. Toupin. Principles of classical mechanics and field theory. In S. Flügge, editor, *Encyclopedia of Physics (Handbuch der Physik)*, volume III/1. Springer, Berlin, 1960.

45. I. Vardoulakis. Private communication, 2005.

46. I. Vardoulakis and E. C. Aifantis. Gradient dependent dilatancy and its implications in shear banding and liquefaction. *Arch. Appl. Mech. (Ingenieur Archiv)*, 59(3):197–208, 1989.

47. J. Weber. Recherches concernant le contraintes intergranulaires dans les milieux pulvérents; application à la rhéologie de ces milieux. *Cahiers français rhéol.*, 2: 161–170, 1966.

48. E. W. Weisstein. Polyhedral Formula. In *http://mathworld.wolfram.com*. Wolfram Research Inc.-CRC Press, 2005.

49. S. N. Wood. Stable and efficient multiple smoothing parameter estimation for generalized additive models. *J. Am. Stat. Assoc.*, 99(467):673–86, 2004.

Generalized Kinetic Maxwell Type Models of Granular Gases

A.V. Bobylev, C. Cercignani, and I.M. Gamba

Summary. In this chapter we consider generalizations of kinetic granular gas models given by Boltzmann equations of Maxwell type. These type of models for non-linear elastic or inelastic interactions, have many applications in physics, dynamics of granular gases, economy, etc. We present the problem and develop its form in the space of characteristic functions, i.e., Fourier transforms of probability measures, from a very general point of view, including those with arbitrary polynomial non-linearities and in any dimension space. We find a whole class of generalized Maxwell models that satisfy properties that characterize the existence and asymptotic of dynamically scaled or self-similar solutions, often referred as *homogeneous cooling states*. Of particular interest is a concept interpreted as an operator generalization of usual Lipschitz conditions which allows to describe the behavior of solutions to the corresponding initial value problem. In particular, we present, in the most general case, existence of self similar solutions and study, in the sense of probability measures, the convergence of dynamically scaled solutions associated with the Cauchy problem to those self-similar solutions, as time goes to infinity. In addition we show that the properties of these self-similar solutions lead to non classical equilibrium stable states exhibiting power tails. These results apply to different specific problems related to the Boltzmann equation (with elastic and inelastic interactions) and show that all physically relevant properties of solutions follow directly from the general theory developed in this presentation.

1 Introduction

It has been noticed in recent years that a significant non-trivial physical phenomena in granular gases can be described mathematically by dissipative Boltzmann type equations, as can be seen in [17] for a review in the area. As motivated by this particular phenomena of energy dissipation at the kinetic level, we consider in this chapter the Boltzmann equation for non-linear interactions of Maxwell type and some generalizations of such models.

The classical conservative (elastic) Boltzmann equation with the Maxwell-type interactions is well-studied in the literature (see [5, 14] and references therein). Roughly speaking, this is a mathematical model of a rarefied gas with binary collisions such that the collision frequency is independent of the velocities of colliding particles, and even though the intermolecular potentials are not of those corresponding to hard sphere interactions, still these models provide a very rich inside to the understanding of kinetic evolution of gases.

Recently, Boltzmann equations of Maxwell type were introduced for models of granular gases were introduced in [7] in three dimensions, and a bit earlier in [3] for in one dimension case. Soon after that, these models became very popular among the community studying granular gases (see, for example, the book [13] and references therein). There are two obvious reasons for such studies The first one is that the inelastic Maxwell–Boltzmann equation can be essentially simplified by the Fourier transform similarly as done for the elastic case, where its study becomes more transparent [6, 7]. The second reason is motivated by the special phenomena associated with *homogeneous cooling* behavior, i.e., solutions to the spatially homogeneous inelastic Maxwell–Boltzmann equation have a non-trivial self-similar asymptotics, and in addition, the corresponding self-similar solution has a power-like tail for large velocities. The latter property was conjectured in [16] and later proved in [9, 11]. This is a rather surprising fact, since the Boltzmann equation for hard spheres inelastic interactions has been shown to have self similar solutions with all moments bounded and large energy tails decaying exponentially. The conjecture of self-similar (or homogeneous cooling) states for such model of Maxwell type interactions was initially based on an exact one-dimensional solution constructed in [1]. It is remarkable that such an asymptotics is absent in the elastic case (as the elastic Boltzmann equation has *too many conservation laws*). Later, the self-similar asymptotics was proved in the elastic case for initial data with infinite energy [8] by using another mathematical tools compared to [9] and [12].

Surprisingly, the recently published exact self-similar solutions [12] for elastic Maxwell type model for a slow down process, derived as a formal asymptotic limit of a mixture, also is shown to have power-like tails. This fact definitely suggests that self-similar asymptotics are related to total energy dissipation rather than local dissipative interactions. As an illustration to this fact, we mention some recent publications [2, 15], where one-dimensional Maxwell-type models were introduced for non-standard applications such as models in economics and social interactions, where also self-similar asymptotics and power-like tail asymptotic states were found.

Thus all the above discussed models describe qualitatively different processes in physics or even in economics, however their solutions have a lot in common from mathematical point of view. It is also clear that some further generalizations are possible: one can, for example, include in the model multiple (not just binary) interactions still assuming the constant

(Maxwell-type) rate of interactions. Will the multi-linear models have similar properties? The answer is *yes*, as we shall see below.

Thus, it becomes clear that there must be some general mathematical properties of Maxwell models, which, in turn, can explain properties of any particular model. That is to say there must be just one *main theorem*, from which one can deduce all above discussed facts and their possible generalizations. Our goal is to consider Maxwell models from very general point of view and to establish their key properties that lead to the self-similar asymptotics.

All the results presented in this chapter are mathematically rigorous. Their full proofs can be found in [10].

After this introduction, we introduce in Sect. 2 three specific Maxwell models of the Boltzmann equation: (A) classical (elastic) Boltzmann equation; (B) the model (A) in the presence of thermostat; (C) inelastic Boltzmann equation for Maxwell type interactions. Then, in Sect. 3, we perform the Fourier transform and introduce an equation that includes all the three models as particular cases. A further generalization is done in Sect. 4, where the concept of generalized multi-linear Maxwell model (in the Fourier space) is introduced. Such models and their generalizations are studied in detail in Sects. 5 and 6. The most important for our approach concept of L-Lipschitz nonlinear operator is explained in Sect. 4. It is shown (Theorem 4.2) that all multi-linear Maxwell models satisfy the L-Lipschitz condition. This property of the models constitutes a basis for the general theory.

The existence and uniqueness of solutions to the initial value problem is stated in Sect. 5.1 (Theorem 5.2). Then, in Sect. 5.2, we present and study the large time asymptotics under very general conditions that are fulfilled, in particular, for all our models. It is shown that *L-Lipschitz* condition leads to self-similar asymptotics, provided the corresponding self-similar solution does exist. The existence and uniqueness of such self-similar solutions is stated in Sect. 5.3 (Theorem 5.12). This theorem can be considered, to some extent, as the *main theorem* for general Maxwell-type models. Then, in Sect. 5.4, we go back to multi-linear models of Sect. 4 and study more specific properties of their self-similar solutions.

We explain in Sect. 6 how to use our theory for applications to any specific model: it is shown that the results can be expressed in terms of just one function $\mu(p)$, $p > 0$, that depends on spectral properties of the specific model. General properties (positivity, power-like tails, etc.) self-similar solutions are studied in Sect. 6.1 and 6.2. It includes also the case of one-dimensional models, where the Laplace (instead of Fourier) transform is used. In Sect. 6.3, we formulate, in the unified statement (Theorem 11.1), the main properties of Maxwell models (A), (B) and (C) of the Boltzmann equation. This result is, in particular, an essential improvement of earlier results of [7] for the model (A) and quite new for the model (B).

Applications to one-dimensional models are also briefly discussed at the end of Sect. 6.3.

2 Maxwell Models of the Boltzmann Equation

We consider a spatially homogeneous rarefied d-dimensional gas ($d = 2, 3, \ldots$) of particles having a unit mass. Let $f(v,t)$, where $v \in \mathbb{R}^d$ and $t \in \mathbb{R}_+$ denote respectively the velocity and time variables, be a one-particle distribution function with usual normalization

$$\int_{\mathbb{R}^d} dv\, f(v,t) = 1. \tag{1}$$

Then $f(v,t)$ has an obvious meaning of a time-dependent probability density in $\mathbb{R}^d$. We assume that the collision frequency is independent of the velocities of the colliding particles (Maxwell-type interactions). We discuss three different physical models (A), (B) and (C).

(A) Classical Maxwell gas (elastic collisions). In this case $f(v,t)$ satisfies the usual Boltzmann equation

$$f_t = Q(f,f) = \int_{\mathbb{R}^d \times S^{d-1}} dw\, d\omega\, g(\frac{u \cdot \omega}{|u|})[f(v')f(w') - f(v)f(w)], \tag{2}$$

where the exchange of velocities after a collision are given by

$$v' = \frac{1}{2}(v + w + |u|\omega), \quad \text{and} \quad w' = \frac{1}{2}(v + w - |u|\omega),$$

where $u = v - w$ is the relative velocity and $\Omega \in S^{d-1}$. For the sake of brevity we shall consider below the model non-negative collision kernels $g(s)$ such that $g(s)$ is integrable on $[-1,1]$. The argument t of $f(v,t)$ and similar functions is often omitted below (as in (2)).

(B) Elastic model with a thermostat. This case corresponds to model (A) in the presence of a thermostat that consists of Maxwell particles with mass $m > 0$ having the Maxwellian distribution

$$M(v) = (\frac{2\pi T}{m})^{-d/2} \exp(-\frac{m|v|^2}{2T}) \tag{3}$$

with a constant temperature $T > 0$. Then the evolution equation for $f(x,t)$ becomes

$$f_t = Q(f,f) + \theta \int dw\, d\omega\, g(\frac{u \cdot \omega}{|u|})[f(v')M(w') - f(v)M(w)], \tag{4}$$

where $\theta > 0$ is a coupling constant, and the exchange of velocities is now

$$v' = \frac{v + m(w + |u|\omega)}{1 + m}, \quad \text{and} \quad w' = \frac{v + mw - |u|\omega}{1 + m},$$

with $u = v - w$ the relative velocity, and $\omega \in S^{d-1}$.

Equation (4) was derived in [12] as a certain limiting case of a binary mixture of weakly interacting Maxwell gases.

(C) Maxwell model for inelastic particles. We consider this model in the form given in [9]. Then the inelastic Boltzmann equation in the weak form reads

$$\frac{\partial}{\partial t}(f, \psi) = \int_{\mathbb{R}^d \times \mathbb{R}^d \times S^{d-1}} dv\, dw\, d\omega\, f(v)f(w)\frac{|u \cdot \omega|}{|u|}[\psi(v') - \psi(v)], \qquad (5)$$

where $\psi(v)$ is a bounded and continuous test function,

$$(f, \psi) = \int_{\mathbb{R}^d} dv\, f(v,t)\psi(v),\ u = v - w,\ \omega \in S^{d-1},\ v' = v - \frac{1+e}{2}(u \cdot \omega)\omega, \quad (6)$$

the constant parameter $0 < e \leq 1$ denotes the restitution coefficient. Note that the model (C) with $e = 1$ is equivalent to the model (A) with some kernel $g(s)$.

All three models can be simplified (in the mathematical sense) by taking the Fourier transform.

We denote

$$\hat{f}(k,t) = \mathcal{F}[f] = (f, e^{-ik \cdot v}), \qquad k \in \mathbb{R}^d, \qquad (7)$$

and obtain (by using the same trick as in [6] for the model (A)) for all three models the following equations:

(A) $$\hat{f}_t = \hat{Q}(\hat{f}, \hat{f}) = \int_{S^{d-1}} d\omega\, g(\frac{k \cdot \omega}{|k|})[\hat{f}(k_+)\hat{f}(k_-) - \hat{f}(k)\hat{f}(0)],$$

$$(8)$$

where $k_\pm = \frac{1}{2}(k \pm |k|\omega)$, $\omega \in S^{d-1}$, $\hat{f}(0) = 1$.

(B) $$\hat{f}_t = \hat{Q}(\hat{f}, \hat{f}) + \theta \int_{S^{d-1}} d\omega\, g(\frac{k \cdot \omega}{|k|})[\hat{f}(k_+)\widehat{M}(k_-) - \hat{f}(k)\widehat{M}(0)],$$

$$(9)$$

where $\widehat{M}(k) = e^{-\frac{T|k|^2}{2m}}$, $k_+ = \frac{k+m|k|\omega}{1+m}$, $k_- = k - k_+$, $\omega \in S^{d-1}$, $\hat{f}(0) = 1$.

(C) $$\hat{f}_t = \int_{S^{d-1}} d\omega\, \frac{|k \cdot \omega|}{|k|}[\hat{f}(k_+)\hat{f}(k_-) - \hat{f}(k)\hat{f}(0)], \qquad (10)$$

where $\hat{f}(0) = 1$, $k_+ = \frac{1+e}{2}(k \cdot \omega)\omega$, $k_- = k - k_+$, with $\omega \in S^{d-1}$ is the direction containing the two centers of the particles at the time of the interaction. Equivalently, one may alternative write $k_- = \frac{1+e}{4}(k - |k|n\tilde{\omega})$, and $k_+ = k - k_-$, where now $\tilde{\omega} \in S^{d-1}$ is the direction of the post collisional relative velocity, and the term $\frac{|k \cdot \omega|}{|k|}\, dw$ is replaced by a function $g(\frac{k \cdot \tilde{\omega}}{|k|})d\tilde{\omega}$.

Case (B) can be simplified by the substitution

$$\hat{f}(k,t) = \tilde{\tilde{f}}(k,t)\exp[-\frac{T|k|^2}{2}], \qquad (11)$$

leading, omitting tildes, to the equation

(B') $$\hat{f}_t = \hat{Q}(\hat{f}, \hat{f}) + \theta \int_{S^{d-1}} d\omega \, g(\frac{k \cdot \omega}{|k|})[\hat{f}(\frac{k + m|k|\omega}{1+m}) - \hat{f}(k)],$$ (12)

i.e., the model for (B) with $T = 0$, or equivalently a linear collisional term the background singular distribution. Therefore, we shall consider below just the case (B'), assuming nevertheless that $\hat{f}(k,t)$ in (12) is the Fourier transform (7) of a probability density $f(v,t)$.

3 Isotropic Maxwell Model in the Fourier Representation

We shall see that these three models (A), (B) and (C) admit a class of isotropic solutions with distribution functions $f = f(|v|,t)$. Indeed, according to (7) we look for solutions $\hat{f} = \hat{f}(|k|,t)$ to the corresponding isotropic Fourier transformed problem, given by

$$x = |k|^2, \quad \varphi(x,t) = \hat{f}(|k|,t) = \mathcal{F}[f(|v|,t)],$$ (13)

where $\varphi(x,t)$ solves the following initial value problem

$$\varphi_t = \int_0^1 ds \, G(s) \left\{\varphi[a(s)x]\varphi[b(s)x] - \varphi(x)\right\} +$$
$$+ \int_0^1 ds H(s) \left\{\varphi[c(s)x] - \varphi(x)\right\},$$ (14)
$$\varphi_{t=0} = \varphi_0(x), \quad \varphi(0,t) = 1,$$

where $a(s)$, $b(s)$, $c(s)$ are non-negative continuous functions on $[0,1]$, whereas $G(s)$ and $H(s)$ are generalized non-negative functions such that

$$\int_0^1 ds \, G(s) < \infty, \quad \int_0^1 ds \, H(s) < \infty.$$ (15)

Thus, we do not exclude such functions as $G = \delta(s - s_0)$, $0 < s_0 < 1$, etc. We shall see below that, for isotropic solutions (13), each of the three equations (8), (10), (12) is a particular case of (14).

Let us first consider (8) with $\hat{f}(k,t) = \varphi(x,t)$ in the notation (13). In that case

$$|k_\pm|^2 = |k|^2 \frac{1 \pm (\omega_0 \cdot \omega)}{2}, \quad \omega_0 = \frac{k}{|k|} \in S^{d-1}, \, d = 2, \ldots,$$

and the integral in (8) reads

$$\int_{S^{d-1}} d\omega \, g(\omega_0 \cdot \omega)\varphi\left[x\frac{1 + \omega_0 \cdot \omega}{2}\right] \varphi\left[x\frac{1 - \omega_0 \cdot \omega}{2}\right].$$ (16)

It is easy to verify the identity

$$\int_{S^{d-1}} d\omega \, F(\omega \cdot \omega_0) = |S^{d-2}| \int_{-1}^{1} dz \, F(z)(1-z^2)^{\frac{d-3}{2}}, \qquad (17)$$

where $|S^{d-2}|$ denotes the "area" of the unit sphere in $\mathbb{R}^{d-1}$ for $d \geq 3$ and $|S^0| = 2$. The identity (17) holds for any function $F(z)$ provided the integral as defined in the right-hand side of (17) exists.

The integral (16) now reads

$$|S^{d-2}| \int_{-1}^{1} dz \, g(z)(1-z^2)^{\frac{d-3}{2}} \varphi(x\frac{1+z}{2})\varphi(x\frac{1-z}{2}) =$$
$$= \int_{0}^{1} ds \, G(s)\varphi(sx)\varphi[(1-s)x],$$

where

$$G(s) = 2^{d-2}|S^{d-2}|g(1-2s)[s(1-s)]^{\frac{d-3}{2}}, \qquad d = 2,3,\dots. \qquad (18)$$

Hence, in this case we obtain (14), where

(A) $\qquad\qquad a(s) - s, \qquad b(s) = 1-s, \qquad H(s) = 0,$

$$(19)$$

$G(s)$ is given in (18).

Two other models (B$'$) and (C), described by (12), (10) respectively, can be considered quite similarly. In both cases we obtain (14), where

(B$'$) $\qquad a(s) = s, \quad b(s) = 1 - s, \quad c(s) = 1 - \dfrac{4m}{(1+m)^2}s,$

$$H(s) = \theta G(s),$$

$$(20)$$

$G(s)$ is given in (18):

(C) $\qquad\qquad a(s) = \dfrac{(1+e)^2}{4}s, \quad b(s) = 1 - \dfrac{(1+e)(3-e)}{4}s,$

$$(21)$$

$$H(s) = 0, \quad G(s) = |S^{d-2}|(1-s)^{\frac{d-3}{2}}.$$

Hence, all three models are described by (14) where $0 < a(s), b(s), c(s) \leq 1$ are non-negative linear functions. One can also find in recent publications some other useful equations that can be reduced after Fourier or Laplace transformations to (14) (see, for example, [2,15] that correspond to the case $G = \delta(s - s_0)$, $H = 0$).

Equation (14) with $H(s) = 0$ first appeared in its general form in [9] in connection with models (A) and (C). The consideration of the problem of self-similar asymptotics for (14) in that paper made it quite clear that the most important properties of "physical" solutions depend very weakly on the specific functions $G(s)$, $a(s)$ and $b(s)$.

4 Models with Multiple Interactions

We present now a general framework to study solutions to the type of problems introduced in the previous section.

We assume, without loss of generality, (scaling transformations $\tilde{t} = \alpha t$, $\alpha = \mathrm{const.}$) that

$$\int_0^1 ds \, [G(s) + H(s)] = 1 \tag{22}$$

in (14). Then (14) can be considered as a particular case of the following equation for a function $u(x,t)$

$$u_t + u = \Gamma(u), \qquad x \geq 0, \ t \geq 0, \tag{23}$$

where

$$\Gamma(u) = \sum_{n=1}^N \alpha_n \Gamma^{(n)}(u), \quad \sum_{n=1}^N \alpha_n = 1, \ \alpha_n \geq 0,$$

$$\Gamma^{(n)}(u) = \int_0^\infty da_1 \dots \int_0^\infty da_n \, A_n(a_1, \dots, a_n) \prod_{k=1}^n u(a_k x), \quad n = 1, \dots, N. \tag{24}$$

We assume that

$$A_n(a) = A_n(a_1, \dots, a_n) \geq 0, \qquad \int_0^\infty da_1 \dots \int_0^\infty da_n \, A(a_1, \dots, a_n) = 1, \tag{25}$$

where $A_n(a) = A_n(a_1, \dots, a_n)$ is a generalized density of a probability measure in $\mathbb{R}_+^n$ for any $n = 1, \dots, N$. We also assume that all $A_n(a)$ have a compact support, i.e.,

$$A_n(a_1, \dots, a_n) \equiv 0 \ \text{if} \ \sum_{k=1}^n a_k^2 > R^2, \qquad n = 1, \dots, N, \tag{26}$$

for sufficiently large $0 < R < \infty$.

Equation (14) is a particular case of (23) with

$$N = 2, \quad \alpha_1 = \int_0^1 ds \, H(s), \quad \alpha_2 = \int_0^1 ds \, G(s)$$

$$A_1(a_1) = \frac{1}{\alpha_1} \int_0^1 ds \, H(s)\delta[a_1 - c(s)] \tag{27}$$

$$A_2(a_1, a_2) = \frac{1}{\alpha_2} \int_0^1 ds \, G(s)\delta[a_1 - a(s)]\delta[a_2 - b(s)].$$

It is clear that (23) can be considered as a generalized Fourier transformed isotropic Maxwell model with multiple interactions provided $u(0,t) = 1$, the case $N = \infty$ in (24) can be treated in the same way.

4.1 Statement of the General Problem

The general problem we consider below can be formulated in the following way. We study the initial value problem

$$u_t + u = \Gamma(u), \quad u_{|t=0} = u_0(x), \quad x \geq 0, \quad t \geq 0, \tag{28}$$

in the Banach space $B = C(\mathbb{R}_+)$ of continuous functions $u(x)$ with the norm

$$\|u\| = \sup_{x>0} |u(x)|. \tag{29}$$

It is usually assumed that $\|u_0\| \leq 1$ and that the operator Γ is given by (24). On the other hand, there are just a few properties of $\Gamma(u)$ that are essential for existence, uniqueness and large time asymptotics of the solution $u(x,t)$ of the problem (28). Therefore, in many cases the results can be applied to more general classes of operators Γ in (28) and more general functional space, for example $B = C(\mathbb{R}^d)$ (anisotropic models). That is why we study below the class (24) of operators Γ as the most important example, but simultaneously indicate which properties of Γ are relevant in each case. In particular, most of the results of Sects. 4–6 do not use a specific form (24) of Γ and, in fact, are valid for a more general class of operators.

Following this way of study, we first consider the problem (28) with Γ given by (24) and point out the most important properties of Γ.

We simplify notations and omit in most of the cases below the argument x of the function $u(x,t)$. The notation $u(t)$ (instead of $u(x,t)$) means then function of the real variable $t \geq 0$ with values in the space $B = C(\mathbb{R}_+)$.

Remark 1. We shall omit below the argument $x \in \mathbb{R}_+$ of functions $u(x)$, $v(x)$, etc., in all cases when this does not cause a misunderstanding. In particular, inequalities of the kind $|u| \leq |v|$ should be understood as a point-wise control in absolute value, i.e., "$|u(x)| \leq |v(x)|$ for any $x \geq 0$" and so on.

We first start by giving the following general definition for operators acting on a unit ball of a Banach space B denoted by

$$U = \{u \in B : \|u\| \leq 1\} \tag{30}$$

Definition 1. *The operator* $\Gamma = \Gamma(u)$ *is called an L-Lipschitz operator if there exists a linear bounded operator* $L : B \to B$ *such that the inequality*

$$|\Gamma(u_1) - \Gamma(u_2)| \leq L(|u_1 - u_2|) \tag{31}$$

holds for any pair of functions $u_{1,2}$ *in* U.

Remark 2. Note that the *L*-Lipschitz condition (31) holds, by definition, at any point $x \in \mathbb{R}_+$ (or $x \in \mathbb{R}^d$ if $B = C(\mathbb{R}^d)$). Thus, condition (31) is much stronger than the classical Lipschitz condition

$$\|\Gamma(u_1) - \Gamma(u_2)\| < C\|u_1 - u_2\| \quad \text{if} \quad u_{1,2} \in U \tag{32}$$

which obviously follows from (31) with the constant $C = \|L\|_B$, the norm of the operator L in the space of bounded operators acting in B. In other words, the terminology "*L*-Lipschitz condition" means the point-wise Lipschitz condition with respect to an specific linear operator L.

We assume, without loss of generality, that the kernels $A_n(a_1, \ldots, a_n)$ in (24) are symmetric with respect to any permutation of the arguments $(a_1, \ldots, a_n)$, $n = 2, 3, \ldots, N$.

The next lemma states that the operator $\Gamma(u)$ defined in (24), which satisfies $\Gamma(1) = 1$ (mass conservation) and maps U into itself, satisfies an *L*-Lipschitz condition, where the linear operator L is the one given by the linearization of Γ near the unity. See [10] for its proof.

Theorem 1. *The operator $\Gamma(u)$ defined in (24) maps U into itself and satisfies the L-Lipschitz condition (31), where the linear operator L is given by*

$$Lu = \int_0^\infty da\, K(a)u(ax), \tag{33}$$

with

$$K(a) = \sum_{n=1}^N n\alpha_n K_n(a), \tag{34}$$

where $K_n(a) = \int_0^\infty da_2 \ldots \int_0^\infty da_n\, A_n(a, a_2, \ldots, a_n)$ *and* $\sum_{n=1}^N \alpha_n = 1.$

for symmetric kernels $A_n(a, a_2, \ldots, a_n), n = 2, \ldots N$.

And the following corollary holds.

Corollary 1. *The Lipschitz condition (32) is fulfilled for $\Gamma(u)$ given in (24) with the constant*

$$C = \|L\| = \sum_{n=1}^N n\alpha_n, \qquad \sum_{n=1}^\infty \alpha_n = 1, \tag{35}$$

where $\|L\|$ is the norm of L in B.

It can also be shown that the *L*-Lipschitz condition holds in $B = C(\mathbb{R}^d)$ for "gain-operators" in Fourier transformed Boltzmann equations (8), (9) and (10).

5 The General Problem in Fourier Representation

5.1 Existence and Uniqueness of Solutions

It is possible to show, with minimal requirements, the existence and uniqueness results associated with the initial value problem (28) in the Banach space $(B, \| \cdot \|)$, where the norm associated to B is defined in (29). In fact, this existence and uniqueness result is an application of the classical Picard iteration scheme and holds for any operator Γ which satisfies the usual Lipschitz condition (32) and transforms the unit ball U into itself. The proof of all statements below can be found in [10].

Lemma 1 (Picard Iteration scheme). *The operator $\Gamma(u)$ maps U into itself and satisfies the L-Lipschitz condition (32), then the initial value problem (28) with arbitrary $u_0 \in U$ has a unique solution $u(t)$ such that $u(t) \in U$ for any $t \geq 0$.*

Next, we observe that the L-Lipschitz condition yields an estimate for the difference of any two solutions of the problems in terms of their initial states. This is a key fact in the development of the further studies of the self similar asymptotics.

Theorem 2. *Consider the Cauchy problem (28) with $\|u_0\| \leq 1$ and assume that the operator $\Gamma : B \to B$*

(a) Maps the closed unit ball $U \subset B$ to itself, and
(b) Satisfies a L-Lipschitz condition (31) for some positive bounded linear operator $L : B \to B$.

Then

(i) There exists a unique solution $u(t)$ of the problem (28) such that $\|u(t)\| \leq 1$ for any $t \geq 0$;
(ii) Any two solutions $u(t)$ and $w(t)$ of problem (28) with initial data in the unit ball U satisfy the inequality

$$|u(t) - w(t)| \leq \exp\{t(L-1)\}(|u_0 - w_0|). \tag{36}$$

Note that under the same conditions as in Theorem 1 the operator Γ given in (24) satisfies necessary conditions for the Theorem 2.

We remind to the reader that the initial value problem (28) appeared as a generalization of the initial value problem (14) for a characteristic function $\varphi(x,t)$, i.e., for the Fourier transform of a probability measure (see (13), (1)). It is important therefore to show that the solution $u(x,t)$ of the problem (28) is a characteristic function for any $t > 0$ provided this is so for $t = 0$, which is addressed in the following statement.

Lemma 2. *Let $U' \subset U \subset B$ be any closed convex subset of the unit ball U (i.e., $u = (1-\theta)u_1 + \theta u_2 \in U'$ for any $u_{1,2} \in U'$ and $\theta \in [0,1]$). If $u_0 \in U'$ in (28) and U is replaced by U' in the condition (1) of Theorem 2, the theorem holds and $u(t) \in U'$ for any $t \geq 0$.*

Remark 3. It is well-known (see, for example, the textbook [18]) that the set $U' \subset U$ of Fourier transforms of probability measures in $\mathbb{R}^d$ (Laplace transforms in the case of $\mathbb{R}_+$) is convex and closed with respect to uniform convergence. On the other hand, it is easy to verify that the inclusion $\Gamma(U') \subset U'$, where Γ is given in (24), holds in both cases of Fourier and Laplace transforms. Hence, all results obtained for (23), (24) can be interpreted in terms of "physical" (positive and satisfying the condition (1)) solutions of corresponding Boltzmann-like equations with multi-linear structure of any order.

We also point out that all results of this section remain valid for operators Γ satisfying conditions (24), with a more general condition such as

$$\sum_{n=1}^{N} \alpha_n \leq 1, \qquad \alpha_n \geq 0, \tag{37}$$

so that $\Gamma(1) < 1$ and so the mass may not be conserved. The only difference in this case is that the operator L satisfying conditions (33), (34) is not a linearization of $\Gamma(u)$ near the unity, but nevertheless Theorem 1 remains true. The inequality (37) is typical for Fourier (Laplace) transformed Smoluchowski-type equations where the total number of particles is decreasing in time (see [21, 22] for related work).

In the next three sections we study in more detail the solutions to the initial value problem (28)–(29) constructed in Theorem 2 and, in particular, their long time behavior, existence, uniqueness, and properties of the self-similar solutions.

5.2 Large Time Asymptotics

The long time asymptotics results are a consequence of some very general properties of operators Γ, namely, that Γ maps the unit ball U of the Banach space $B = C(\mathbb{R}_+)$ into itself, Γ is an L-Lipschitz operator (i.e., satisfies (31)) and that Γ is invariant under dilations.

These three properties are sufficient to study self-similar solutions and large time asymptotic behavior for the solution to the Cauchy problem (28) in the unit ball U of the Banach space $C(\mathbb{R}_+)$.

Main properties of the operator Γ :

(a) Γ maps the unit ball U of the Banach space $B = C(\mathbb{R}_+)$ into itself, that is

$$\|\Gamma(u)\| \leq 1 \quad \text{for any} \quad u \in C(\mathbb{R}_+) \quad \text{such that} \quad \|u\| \leq 1. \tag{38}$$

(b) Γ is an L-Lipschitz operator (i.e., satisfies (31)) with L from (33), i.e.,

$$|\Gamma(u_1) - \Gamma(u_2)|(x) \leq L(|u_1 - u_2|)(x) = \int_0^\infty da K(a)|u_1(ax) - u_2(ax)|, \tag{39}$$

for $K(a) \geq 0$, for all $x \geq 0$ and for any two functions $u_{1,2} \in C(\mathbb{R}_+)$ such that $\|u_{1,2}\| \leq 1$.

(c) Γ is invariant under dilations:

$$e^{\tau \mathcal{D}} \Gamma(u) = \Gamma(e^{\tau \mathcal{D}} u), \quad \mathcal{D} = x\frac{\partial}{\partial x}, \quad e^{\tau \mathcal{D}} u(x) = u(xe^\tau), \quad \tau \in \mathbb{R}. \tag{40}$$

No specific information about Γ beyond these three conditions will be used in this section.

It was already shown in Theorem 2 that the conditions (a) and (b) guarantee existence and uniqueness of the solution $u(x,t)$ to the initial value problem (28)–(29). The property (b) yields the estimate (36) that is very important for large time asymptotics, as we shall see below. The property (c) suggests a special class of self-similar solutions to (28).

We recall the usual meaning of the notation $y = O(x^p)$ (often used below): $y = O(x^p)$ if and only if there exists a positive constant C such that

$$|y(x)| \leq Cx^p \quad \text{for any} \quad x \geq 0. \tag{41}$$

In order to study long time stability properties to solutions whose initial data differs in terms of $O(x^p)$, we will need some spectral properties of the linear operator L.

Definition 2. *Let L be the positive linear operator given in (33), (34), then*

$$Lx^p = \lambda(p)x^p, \qquad 0 < \lambda(p) = \int_0^\infty da\, K(a)a^p < \infty, \quad p \geq 0, \tag{42}$$

and the spectral function $\mu(p)$ is defined by

$$\mu(p) = \frac{\lambda(p) - 1}{p}. \tag{43}$$

An immediate consequence of properties (a) and (b), as stated in (39), is that one can obtain a criterion for a point-wise in x estimate of the difference of two solutions to the initial value problem (28) yielding decay properties depending on the spectrum of L, as the following statement and its corollary assert.

Lemma 3. *Let $u_{1,2}(x,t)$ be any two classical solutions of the problem* (28) *with initial data satisfying the conditions*

$$|u_{1,2}(x,0)| \leq 1, \qquad |u_1(x,0) - u_2(x,0)| \leq C\, x^p, \quad x \geq 0 \tag{44}$$

for some positive constant C and p. Then

$$|u_1(x,t) - u_2(x,t)| \leq C x^p\, e^{-t(1-\lambda(p))}, \qquad \text{for all } t \geq 0 \tag{45}$$

Corollary 2. *The minimal constant C for which condition* (44) *is satisfied is*

$$C_0 = \sup_{x \geq 0} \frac{|u_1(x,0) - u_2(x,0)|}{x^p} = \left\| \frac{u_1(x,0) - u_2(x,0)}{x^p} \right\|, \tag{46}$$

and the following estimate holds

$$\left\| \frac{u_1(x,t) - u_2(x,t)}{x^p} \right\| \leq e^{-t(1-\lambda(p))} \left\| \frac{u_1(x,0) - u_2(x,0)}{x^p} \right\| \tag{47}$$

for any $p > 0$.

A result similar to Lemma 3 was first obtained in [9] for the inelastic Boltzmann equation whose Fourier transform is given in example (C), (10). Its corollary in the form similar to (47) for (10) was stated later in [11] and was interpreted there as "the contraction property of the Boltzmann operator" (note that the left-hand side of (47) can be understood as a distance between two solutions). Independently of the terminology. the key reason for estimates (45)–(47) is the Lipschitz property of the operator Γ. It is remarkable that the large time asymptotics of $u(x,t)$, satisfying the problem (28) with such Γ, can be explicitly expressed through spectral characteristics of the linear operator L.

In order to study the large time asymptotics of $u(x,t)$ in more detail we distinguish two different kinds of asymptotic behavior:

(1) Convergence to stationary solutions
(2) Convergence to self-similar solutions provided the condition (c), of the main properties on Γ, is satisfied

The case (1) is relatively simple. Any stationary solution $\bar{u}(x)$ of the problem (28) satisfies the equation

$$\Gamma(\bar{u}) = \bar{u}, \qquad \bar{u} \in C(\mathbb{R}_+), \quad \|\bar{u}\| \leq 1. \tag{48}$$

If the stationary solution $\bar{u}(x)$ does exists (note, for example, that $\Gamma(0) = 0$ and $\Gamma(1) = 1$ for Γ given in (24)) then the large time asymptotics of some classes of initial data $u_0(x)$ in (28) can be studied directly on the basis of Lemma 3. It is enough to assume that $|u_0(x) - \bar{u}(x)|$ satisfies (44) with p such that $\lambda(p) < 1$. Then $u(x,t) \to \bar{u}(x)$ as $t \to \infty$, for any $x \geq 0$.

This simple consideration, however, does not answer at least two questions:

(A) What happens with $u(x,t)$ if the inequality (44) for $|u_0(x) - \bar{u}(x)|$ is satisfied with such p that $\lambda(p) > 1$?

(B) What happens with $u(x,t)$ for large x (note that the estimate (45) becomes trivial if $x \to \infty$).

In order to address these questions we consider a special class of solutions of (28), the so-called self-similar solutions. Indeed the property (c) of Γ shows that (28) admits a class of formal solutions $u_s(x,t) = w(x\,e^{\mu_* t})$ with some real μ_*. It is convenient for our goals to use a terminology that slightly differs from the usual one.

Definition 3. *The function $w(x)$ is called a self-similar solution associated with the initial value problem* (28) *if it satisfies the problem*

$$\mu_* \mathcal{D}w + w = \Gamma(w), \qquad \|w\| \le 1, \tag{49}$$

in the notation of (40), (24).

The convergence of solutions $u(x,t)$ of the initial value problem (28) to a stationary solution $\bar{u}(x)$ can be considered as a special case of the self-similar asymptotics with $\mu_* = 0$.

Under the assumption that self-similar solutions exists (the existence is proved in the next section), we state the fundamental result on the convergence of solutions $u(x,t)$ of the initial value problem (28) to self-similar ones (sometimes called in the literature *self-similar stability*).

Lemma 4. *We assume that*

(i) *For some $\mu_* \in \mathbb{R}$, there exists a classical (continuously differentiable if $\mu_* \ne 0$) solution $w(x)$ of* (49) *such that $\|w\| \le 1$;*

(ii) *The initial data $u(x,0) = u_0$ in the problem* (28) *satisfies*

$$u_0 = w + O(x^p), \qquad \|u_0\| \le 1, \;\; for \;\; p > 0 \;\; such \; that \; \mu(p) < \mu_*, \tag{50}$$

where $\mu(p)$ defined in (43) *is the spectral function associated to the operator L.*
Then

$$|u(xe^{-\mu_* t}, t) - w(x)| = O(x^p)e^{-pt(\mu_* - \mu(p))} \tag{51}$$

and therefore

$$\lim_{t \to \infty} u(xe^{-\mu_* t}, t) = w(x), \qquad x \ge 0. \tag{52}$$

Remark 4. Lemma 4 shows how to find a domain of attraction of any self-similar solution provided the self-similar solution is itself known. It is remarkable that the domain of attraction can be expressed in terms of just the spectral function $\mu(p)$, $p > 0$, defined in (43), associated with the linear operator L for which the operator Γ satisfies the L-Lipschitz condition.

Generally speaking, the equality (52) can be also fulfilled for some other values of p with $\mu(p) > \mu_*$ in (50), but, at least, it always holds if $\mu(p) < \mu_*$.

We shall need some properties of the *spectral function* $\mu(p)$. Having in mind further applications, we formulate these properties in terms of the operator Γ given in (24), though they depend only on $K(a)$ in (42)

Lemma 5. *The spectral function $\mu(p)$ has the following properties:*

(i) It is positive and unbounded as $p \to 0^+$, with asymptotic behavior given by

$$\mu(p) \approx \frac{\lambda(0) - 1}{p}, \qquad p \to 0, \tag{53}$$

where, for Γ from (24)

$$\lambda(0) = \int_0^\infty da\, K(a) = \sum_{n=1}^N \alpha_n n \geq 1, \qquad \sum_{n=1}^N \alpha_n = 1,\ \alpha_n \geq 0, \tag{54}$$

and therefore $\lambda(0) = 1$ if and only if the operator Γ (24) is linear ($N = 1$); (ii) In the case of a multi-linear Γ operator, there is not more than one point $0 < p_0 < \infty$, where the spectral function $\mu(p)$ achieves its minimum, that is, $\mu'(p_0) = \frac{d\mu}{dp}(p_0) = 0$, with $\mu(p_0) \leq \mu(p)$ for any $p > 0$, provided $N \geq 2$ and $\alpha_N > 0$.

Remark 5. From now on, we shall always assume below that the operator Γ from (24) is multi-linear. Otherwise it is easy to see that the problem (49) has no solutions (the condition $\|w\| \leq 1$ is important!) except for the trivial ones $w = 0, 1$.

The following corollaries are readily obtained from Lemma 5 part (ii) and its proof.

Corollary 3. *For the case of a non-linear Γ operator, i.e., $N \geq 2$, the spectral function $\mu(p)$ is always monotone decreasing in the interval $(0, p_0)$, and $\mu(p) \geq \mu(p_0)$ for $0 < p < p_0$. This implies that there exists a unique inverse function $p(\mu) : (\mu(p_0), +\infty) \to (0, p_0)$, monotone decreasing in its domain of definition.*

Corollary 4. *There are precisely four different kinds of qualitative behavior of $\mu(p)$ shown on Fig. 1.*

Proof. There are two options: $\mu(p)$ is monotone decreasing function (Fig. 1a) or $\mu(p)$ has a minimum at $p = p_0$ (Fig. 1b–d). In case Fig. 1a $\mu(p) > 0$ for all $p > 0$ since $\mu(p) > 1/p$. The asymptotics of $\lambda(p)$ (42) is clear:

$$(1) \qquad \lambda(p) \xrightarrow[p \to \infty]{} \lambda_\infty \in \mathbb{R}_+ \quad \text{if} \quad \int_{1+}^\infty da\, K(a) = 0; \tag{55}$$

$$(2) \qquad \lambda(p) \xrightarrow[p \to \infty]{} \infty \quad \text{if} \quad \int_{1+}^\infty da\, K(a) > 0. \tag{56}$$

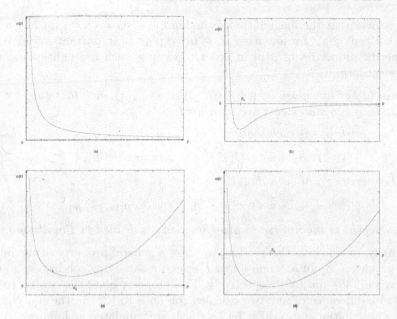

Fig. 1. Possible profiles of the spectral function $\mu(p)$

In the case (1) when $\mu(p) \to \infty$ as $p \to 0$, two possible pictures (with and without minimum) are shown on Fig. 1b and Fig. 1a, respectively. In case (2), from (42) it is clear that $\lambda(p)$ grows exponentially for large p, therefore $\mu(p) \to \infty$ as $p \to \infty$. Then the minimum always exists and we can distinguish two cases: $\mu(p_0) < 0$ (Fig. 1d) and $\mu(p_0) > 0$ (Fig. 1c).

We note that, for Maxwell models (A), (B), (C) of Boltzmann equation (Sects. 2 and 3), only cases (a) and (b) of Fig. 1 can be possible (actually this is the case (b)) since the condition (55) holds. Figure 1 gives a clear graphic representation of the domains of attraction of self-similar solutions (Lemma 4): it is sufficient to draw the line $\mu(p) = \mu_* = constant$, and to consider a p such that the graph of $\mu(p)$ lies below this line.

Therefore, the following corollary follows directly from the properties of the spectral function $\mu(p)$, as characterized by the behaviors in Fig. 1, where we assume that $\mu(p_0) = 0$ for $p_0 = \infty$, for the case shown on Fig. 1a.

Corollary 5. *Any self-similar solution* $u_s(x,t) = w(xe^{\mu_* t})$ *with* $\mu(p_0) < \mu_* < \infty$ *has a non-empty domain of attraction, where* p_0 *is the unique (minimum) critical point of the spectral function* $\mu(p)$.

Proof. We use Lemma 4 part (ii) on any initial state $u_0 = w + O(x^p)$ with $p > 0$ such that $\mu(p_0) \le \mu(p) < \mu_*$. In particular, (51) and (52) show that the domain of attraction of $w(xe^{\mu_* t})$ contains any solution to the initial value problem (28) with the initial state as above.

It is clear that the inequalities of the kind $u_1 - u_2 = O(x^p)$ for any $p > 0$ such that $\mu(p) < \mu_*$, for any fixed $\mu_* \geq \mu(p_0)$ play an important role. We can use specific properties of $\mu(p)$ in order to express such inequalities in more convenient form.

Lemma 6. *For any given $\mu_* \in (\mu(p_0), \infty)$ and $u_{1,2}(x)$ such that $\|u_{1,2}\| < \infty$, the following two statements are equivalent:*

(i) There exists $p > 0$ such that

$$u_1 - u_2 = O(x^p), \qquad \text{with} \quad \mu(p) < \mu_*. \tag{57}$$

(ii) There exists $\varepsilon > 0$ such that

$$u_1 - u_2 = O(x^{\mathrm{p}(\mu_*)+\varepsilon}), \qquad \text{with} \quad p(\mu_*) < p_0, \tag{58}$$

where $p(\mu)$ is the inverse to $\mu(p)$ function, as defined in Corollary 3.

Finally, to conclude this section, we show a general property of the initial value problem (28) for any non-linear Γ operator satisfying conditions (a) and (b) given in (38) and (39) respectively. This property gives the control to the point-wise difference of any two rescaled solutions to (28) in the unit sphere of B , whose initial states differ by $O(x^p)$. It is formulated as follows.

Lemma 7. *Consider the problem* (28), *where Γ satisfies the conditions (a) and (b). Let $u_{1,2}(x,t)$ are two solutions satisfying the initial conditions $u_{1,2}(x,0) = u_0^{1,2}(x)$ such that*

$$\|u_0^{1,2}\| \leq 1, \quad u_0^1 - u_0^2 = O(x^p), \quad p > 0. \tag{59}$$

then, for any real μ_*,

$$\Delta_{\mu_*}(x,t) = u_1(xe^{-\mu_* t}, t) - u_2(xe^{-\mu_* t}, t) = O(x^p)e^{-pt[\mu_* - \mu(p)]} \tag{60}$$

and therefore

$$\lim_{t \to \infty} \Delta_{\mu_*}(x,t) = 0, \qquad x \geq 0, \tag{61}$$

for any $\mu_* > \mu(p)$.

Remark 6. There is an important point to understand here. Lemmas 3 and 4 hold for any operator Γ that satisfies just the two properties (a) and (b) stated in (38) and (39). It says that, in some sense, a distance between any two solutions with initial conditions satisfying (59) tends to zero as $t \to \infty$. Such terminology and corresponding distances were introduced for specific forms of Maxwell–Boltzmann models in [4, 19]. It should be pointed out, however, that this *contraction property* may not say much about large time asymptotics of $u(x,t)$, unless the corresponding self-similar solutions are known, for which the operator Γ must be invariant under dilations (so it satisfies also property (c) as well, as stated in (40)). In such case one can use estimate (61) to deduce the convergence in the form (52), (53).

Therefore one must study the problem of existence of self-similar solutions, which is considered in the next section.

5.3 Existence of Self-Similar Solutions

We develop now a criteria for existence, uniqueness and self-similar asymptotics to the problem (49) for any operator Γ that satisfies conditions (a), (b) and (c) from Sect. 6, with the corresponding spectral function $\mu(p)$ defined in (43).

Theorem 3 below shows the criteria for existence and uniqueness of self-similar solutions for any operator Γ that satisfies just conditions (a) and (b). Then Theorem 4 follows, showing a general criteria to self-similar asymptotics for the problem (28) for any operator Γ that satisfying conditions (a), (b) and (c).

We consider (49) written in the form

$$\mu x w'(x) + w(x) = g(x), \qquad g = \Gamma(w), \ \mu \in \mathbb{R}, \tag{62}$$

and, assuming that $\|w\| < \infty$, transform this equation to the integral form. It is easy to verify that the resulting integral equation reads

$$w(x) = \int_0^1 d\tau \, g(x\tau^\mu). \tag{63}$$

By means of an iteration scheme, the following result can be proved.

Theorem 3. *Consider* (62) *with arbitrary* $\mu \in \mathbb{R}$ *and the operator* Γ *satisfying the conditions (a) and (b) from Sect. 6. Assume that there exists a continuous function* $w_0(x)$, $x \geq 0$, *such that*

(i) $\|w_0\| \leq 1$ *and*

(ii)

$$\int_0^1 d\tau \, g_0(x\tau^\mu) = w_0(x) + O(x^p), \qquad g_0 = \Gamma(w_0), \tag{64}$$

with some $p > 0$ *satisfying the inequality*

$$\mu(p) = \frac{1}{p}[\int_0^\infty da \, K(a)a^p - 1] < \mu. \tag{65}$$

Then there exists a classical solution $w(x)$ *of* (62). *The solution is unique in the class of continuous functions satisfying conditions*

$$\|w\| \leq 1, \qquad w(x) = w_0(x) + O(x^{p_1}), \tag{66}$$

with any p_1 *such that* $\mu(p_1) < \mu$.

Combining Lemmas 1 and 4, the following general statement related to the self-similar asymptotics for the problem (28) is obtained.

Theorem 4. *Let $u(x,t)$ be a solution of the problem (28) with $\|u_0\| \leq 1$ and Γ satisfying the conditions (a), (b), (c) from Sect. 6. Let $\mu(p)$ denote the spectral function (65) having its minimum (infimum) at $p = p_0$ (see Fig. 1), the case $p_0 = \infty$ is also included. We assume that there exists $p \in (0, p_0)$ and $0 < \varepsilon < p_0 - p$ such that*

$$\int_0^1 d\tau \, g_0(x\tau^{\mu(p)}) = u_0(x) + 0(x^{p+\varepsilon}), \qquad g_0 = \Gamma(u_0), \quad \varepsilon > 0. \qquad (67)$$

Then

(i) There exists a unique solution $w(x)$ of (62) with $\mu = \mu(p)$ such that

$$\|w\| \leq 1, \qquad w(x) = u_0(x) + O(x^{p+\varepsilon}), \qquad (68)$$

(ii)

$$\lim_{t \to \infty} u(x \, e^{-\mu(p)t}, t) = w(x), \qquad x \geq 0, \qquad (69)$$

where the convergence is uniform on any bounded interval in $\mathbb{R}_+$ and

$$u(x \, e^{-\mu(p)t}, t) - w(x) = O(x^{p+\varepsilon} e^{-\beta(p,\varepsilon)t}), \qquad (70)$$

with $\beta(p, \varepsilon) = (p + \varepsilon)(\mu(p) - \mu(p + \varepsilon)) > 0$.

Hence, a general criterion (68) is obtained for the self-similar asymptotics of $u(x,t)$ with a given initial condition $u_0(x)$. The criterion can be applied to the problem (28) with any operator Γ satisfying conditions (a), (b), (c) from Sect. 5.2. The specific class (24) of operators Γ is studied in Sect. 6. We shall see below that the condition (68) can be essentially simplified for such operators.

5.4 Properties of Self-Similar Solutions

We now apply the general theory (in particular, Theorem 4) to the particular case of the multi-linear operators Γ considered in Sect. 4, where their corresponding spectral function $\mu(p)$ satisfies (65), (34) whose behavior corresponds to Fig. 1. We also show that $p_0 = \min_{p>0} \mu(p) > 1$ is a necessary condition for self-similar asymptotics.

In addition, Theorem 5 establish sufficient conditions for which self-similar solutions of problem (62) will lead to well defined self-similar solutions (distribution functions) of the original problem after taking the inverse Fourier transform.

We consider the integral equation (63) written as

$$w = \Gamma_\mu(w) = \int_0^1 dt \, g(xt^\mu), \qquad g = \Gamma(w), \, \mu \in \mathbb{R}. \qquad (71)$$

The following two properties of $w(x)$ that are independent of the specific form (24) of Γ.

Lemma 8.

(i) If there exist a closed subset $U' \subset U$ of the unit ball U in B, such that $\Gamma_\mu(U') \subset U'$ for any $\mu \in \mathbb{R}$, and for some function $w_0 \in U'$ the conditions of Theorem 3 are satisfied, then $w \in U'$.

(ii) If the conditions of Theorem 3 for Γ are satisfied and, in addition, $\Gamma(1) = 1$, then the solution $w_* = 1$ of (71) is unique in the class of functions $w(x)$ satisfying the condition

$$w(x) = 1 + O(x^p), \qquad \mu(p) < \mu. \tag{72}$$

We observe that the statement (ii) can be interpreted as a necessary condition for existence of non-trivial ($w \neq$ const.) solutions of (71): if there exists a non-trivial solution $w(x)$ of (71), where $\Gamma(1) = 1$, such that

$$\|w\| = 1, \qquad w = 1 + O(x^p), \quad p > 0, \quad \text{then} \quad \mu \leq \mu(p). \tag{73}$$

We recall that $\mu(p)$ satisfies the inequality $\mu(p) \geq \mu(p_0)$ (Fig. 1).

If $p \geq p_0$ (provided $p_0 < \infty$) in (73), then there are no non-trivial solutions with $\mu > \mu(p_0)$.

On the other hand, possible solutions with $\mu \leq \mu(p_0)$ (even if they exist) are irrelevant for the problem (28) since they have an empty domain of attraction (Lemma 4).

Therefore we always assume below that $\mu > \mu(p_0)$ and, consequently, $p \in (0, p_0)$ in (73).

Let us consider now the specific class (24)–(25) of operators Γ, with functions $u(x)$ satisfying the condition $u(0) = 1$. Then, $u(0, t) = 1$ for the solution $u(x, t)$ of the problem (28).

In addition, the operators (24) are invariant under dilation transformations (40) (property (c), Sect. 6). Therefore, the problem (28) with the initial condition $u_0(x)$ satisfying

$$u(0) = 1, \quad \|u_0\| = 1; \qquad u_0(x) = 1 - \beta x^p + \cdots, \quad x \to 0, \tag{74}$$

can be always reduced to the case $\beta = 1$ by transformation $x' = x\beta^{1/p}$.

Moreover, the whole class of operators (24) with different kernels $A_n(a_1, \ldots, a_n)$, $n = 1, 2, \ldots$, is invariant under transformations $\tilde{x} = x^p$, $p > 0$. The result of such transformation of Γ is another operator $\tilde{\Gamma}$ of the same class (24) with kernels $\tilde{A}_n(a_1, \ldots, a_n)$.

Therefore we fix the initial condition (74) with $\beta = 1$ and transform the function (74) and the (28) to new variables $\tilde{x} = x^p$. Then we omit the tildes and reduce the problem (28), with initial condition (74) to the case $\beta = 1$, $p = 1$. We study this case in detail and formulate afterward the results in terms of initial variables.

Next, we assume a bit more about the asymptotics of the initial data $u_0(x)$ for small x, namely

$$\|u_0\| = 1, \quad u_0(x) = 1 - x + O(x^{1+\varepsilon}), \quad x \to 0, \tag{75}$$

with some $\varepsilon > 0$.

Then, our goal now is to apply the general theory (in particular, Theorem 4 and criterion (67)) to this particular case. We assume that the spectral function $\mu(p)$ given by (65), (34), corresponds to one of the four cases shown on Fig. 1 with $p_0 > 1$.

Let us take a typical function $u_0 = e^{-x}$ satisfying (75) and apply the criterion (67), from Theorem 4 or, equivalently, look for such $p > 0$ that (67) is satisfied. That is, find possibles values of $p > 0$ such that

$$\Gamma_{\mu(p)}(e^{-x}) - e^{-x} = 0(x^{p+\varepsilon}), \tag{76}$$

in the notation of (71).

It is important to observe that now the spectral function $\mu(p)$ is closely connected with the operator Γ (see (65) and (34)), since this was not assumed in the general theory of Sects. 4–7. This connection leads to much more specific results, than, for example, the general Theorems 3, and 4.

The properties of self-similar solutions to problem (62) for $p_0 > 1$, and consequently, for

$$\mu(p) \geq \mu(p_0) > -\frac{1}{p_0} > -1, \tag{77}$$

can be obtained from the structure of $\Gamma_\mu(e^{-x})$ for any $\mu > -1$ using its explicit formula ((65) and (34)). In particular, for

$$\Gamma_\mu(e^{-x}) = \sum_{n=1}^{N} \alpha_n \int_{\mathbb{R}_+^n} da_1 \ldots da_n A_n(a_1, \ldots, a_n) I_\mu\left[x \sum_{k=1}^{n} a_k\right], \tag{78}$$

where

$$I_\mu(y) = \int_0^1 dt\, e^{-yt^\mu}, \quad \mu \in \mathbb{R},\ y > 0, \qquad \sum_{n=1}^{N} \alpha_n = 1. \tag{79}$$

the following two statements can be proven.

Lemma 9. *The condition (76) is fulfilled if and only if $p \leq 1$, and therefore $\mu(p) \geq \mu(1)$ whenever $p_0 > 1$ with $\mu(p_0) = \min_{p>0} \mu(p)$.*

Theorem 5. *The limiting function $w(x)$ constructed by the iteration scheme to prove existence in the solution w in Theorem 3 satisfies (71) with $\mu = \mu(1)$, where Γ_- is given in (24), $\mu(p)$ is defined in (65), (34). Then, the following conditions are fulfilled for $w(x)$:*

(1) It satisfies

$$0 \leq\, < w(x) \leq 1, \qquad with\ w(0) = 1\ \ and\ \ w'(0) = -1, \qquad (80)$$

$$w'(x) \leq 0, \qquad |w'(x)| \leq 1, \qquad and \qquad w(x) = e^{-x} + O(x^{\pi(\mu)}), \qquad (81)$$

with

$$\pi(\mu) = \begin{cases} 2 & if\ \mu > -\frac{1}{2}, \\ 2 - \varepsilon & for\ any\ \varepsilon > 0\ \ if\ \mu = \frac{1}{2}, \\ \frac{1}{|\mu|} & if\ -1 < \mu < -\frac{1}{2}. \end{cases}$$

(2) Further

$$e^{-x} \leq w(x) \leq 1, \qquad \lim_{x \to \infty} w(x) = 0, \qquad and \qquad (82)$$

(3) There exists a generalized non-negative function $R(\tau)$, $\tau \geq 0$, such that

$$w(x) = \int_0^\infty d\tau\, R(\tau) e^{-\tau x} \qquad \int_0^\infty d\tau\, R(\tau) = \int_0^\infty d\tau\, R(\tau)\tau = 1. \quad (83)$$

The integral representation (83) is important for properties of corresponding distribution functions satisfying Boltzmann-type equations. Now it is easy to return to initial variables with u_0 given in (74) and to describe the complete picture of the self-similar relaxation for the problem (28).

6 Main Results for Maxwell Models with Multiple Interactions

6.1 Self-Similar Asymptotics

We apply now the results of Sect. 6 to the specific case when the Cauchy problem (28) with a fixed operator Γ (24) corresponds to the Fourier transform problem for Maxwell models with multiple interactions. In particular we study the time evolution of $u_0(x)$ satisfying the conditions

$$\|u_0\| = 1; \quad u_0 = 1 - x^p + O(x^{p+\varepsilon}), \ x \to 0, \qquad (84)$$

with some positive p and ε. Then, from Theorems 3 and 4, there exists a unique classical solution $u(x,t)$ of the problem (28), (84) such that, for all $t \geq 0$,

$$\|u(\cdot, t)\| = 1; \qquad u(x, t) = 1 + O(x^p), \ x \to 0. \qquad (85)$$

First, consider the linearized operator L given in (33)–(34) and construct the spectral function $\mu(p)$ given in (65) which will be of one of four kinds described qualitatively on Fig. 1.

Second, find the value $p_0 > 0$ that corresponds to minimum (infimum) of $\mu(p)$. Note that $p_0 = \infty$ just for the case described on Fig. 1a, otherwise $0 < p_0 < \infty$. Compare p_0 with the value p from (84). If $p < p_0$ then the problem (28), (84) has a self-similar asymptotics (see below).

In particular, two different cases are possible: (1) $p \geq p_0$ provided $p_0 < \infty$; (2) $0 < p < p_0$. In the first case a behavior of $u(x,t)$ for large t may depend strictly on initial conditions.

Depending on how p compares with p_0, we can obtain. We again use Lemma 7 with $u_1 = u$ and $u_2 = u_s = \psi(xe^{\mu(p)t})$ and obtain for the solution $u(x,t)$ of the problem (28), (84):

$$\lim_{t \to \infty} u(xe^{-\mu t}, t) = \begin{cases} 1 & \text{if } \mu > \mu(p) \\ \psi(x) & \text{if } \mu = \mu(p) \\ 0 & \text{if } \mu(p) > \mu > \mu(p + \delta), \end{cases} \tag{86}$$

with sufficiently small $\delta > 0$.

We see that $\psi(x) = w(x^p)$, where $w(x)$ has all properties described in Theorem 5. The equalities (86) explain the exact meaning of the approximate identity,

$$u(x,t) \approx \psi(xe^{\mu(p)t}), \qquad t \to \infty, \quad xe^{\mu(p)t} = \text{const.}, \tag{87}$$

that we call self-similar asymptotics.

In particular, the following statement holds.

Proposition 1. *The solution $u(x,t)$ of the problem (28), (84), with Γ given in (24), satisfies either one of the following limiting identities:*

(1)

$$\lim_{t \to \infty} u(xe^{-\mu t}, t) = 1, \qquad x \geq 0, \tag{88}$$

for any $\mu > \mu(p_0)$. if $p \geq p_0$ for the initial data (84),
(2) Equation (86) provided $0 < p < p_0$.

The convergence in (88), (86) is uniform on any bounded interval $0 \leq x \leq R$, and

$$u(xe^{\mu(p)t}, t) - \psi(x) = O(x^{p+\varepsilon})e^{-\beta(p,\varepsilon)t}, \qquad \beta(p,\varepsilon) = (p+\varepsilon)(\mu(p) - \mu(p+\varepsilon)),$$

for $0 < p < p_0$ and $0 < \varepsilon < p_0 - p$.

Remark 7. There is a connection between self-similar asymptotics and non-linear wave propagation. It is easy to see that self-similar asymptotics becomes more transparent in logaritheoremic variables

$$y = \ln x, \quad u(x,t) = \hat{u}(y,t), \quad \psi(x,t) = \hat{\psi}(y,t)$$

Thus, (87) becomes

$$\hat{u}(y,t) \approx \hat{\psi}(y + \mu(p)t), \qquad t \to \infty, \quad y + \mu(p)t = \text{const.}, \tag{89}$$

hence, the self-similar solutions are simply nonlinear waves (note that $\psi(-\infty) = 1$, $\psi(+\infty) = 0$) propagating with constant velocities $c_p = -\mu(p)$ to the right if $c_p > 0$ or to the left if $c_p < 0$. If $c_p > 0$ then the value $u(-\infty, t) = 1$

is transported to any given point $y \in \mathbb{R}$ when $t \to \infty$. If $c_p < 0$ then the profile of the wave looks more naturally for the functions $\tilde{u} = 1 - \hat{u}$, $\tilde{\psi} = 1 - \psi$.

We conclude that (28) can be considered in some sense as the equation for nonlinear waves. The self-similar asymptotics (89) means a formation of the traveling wave with a universal profile for a broad class of initial conditions. This is a purely non-linear phenomenon, it is easy to see that such asymptotics cannot occur in the particular case ($N = 1$ in (24)) of the linear operator Γ.

6.2 Distribution Functions, Moments and Power-Like Tails

We have described above the general picture of behavior of the solutions $u(x,t)$ to the problem (28), (84). On the other hand, (28) (in particular, its special case (14)) was obtained as the Fourier transform of the kinetic equation. Therefore we need to study in more detail the corresponding distribution functions.

Set $u_0(x)$ in the problem (28) to be an isotropic characteristic function of a probability measure in $\mathbb{R}^d$, i.e.,

$$u_0(x) = \mathcal{F}[f_0] = \int_{\mathbb{R}^d} dv\, f_0(|v|)e^{-ik \cdot v}, \qquad k \in \mathbb{R}^d,\; x = |k|^2, \qquad (90)$$

where f_0 is a generalized positive function normalized such that $u_0(0) = 1$ (distribution function). Let U be a closed unit ball in the $B = C(\mathbb{R}_+)$ as defined in (30).

Then, we can apply all results of Sect. 5, and conclude that there exists a distribution function $f(v,t)$, $v \in \mathbb{R}^d$, satisfying (1), such that

$$u(x,t) = \mathcal{F}[f(\cdot,t)], \qquad x = |k|^2, \qquad (91)$$

for any $t \geq 0$, and a similar conclusion can be obtain if we assume the Laplace (instead of Fourier) transform in (90).

Then there exists a distribution function $f(v,t)$, $v > 0$, such that

$$u(x,t) = \mathcal{L}[f(\cdot,t)] = \int_0^\infty dv\, f(v,t)e^{-xv}, \qquad u(0,t) = 1,\; x \geq 0,\; t \geq 0,\; (92)$$

where $u(x,t)$ is the solution of the problem (28) constructed in Theorem 2 and Lemma 2.

The approximate equation (87) in terms of distribution functions (91) reads

$$f(|v|,t) \simeq e^{-\frac{d}{2}\mu(p)\,t} F_p(|v|e^{-\frac{1}{2}\mu(p)\,t}), \qquad t \to \infty,\; |v|e^{-\frac{1}{2}\mu(p)\,t} = \text{const.}, \qquad (93)$$

where $F_p(|v|)$ is a distribution function such that

$$\psi_p(x) = \mathcal{F}[F_p], \qquad x = |k|^2, \qquad (94)$$

with ψ_p given by

$$u_s(x,t) = \psi(xe^{\mu(p)t}) \tag{95}$$

(the notation ψ_p is used in order to stress that ψ defined in (95), depends on p). The factor $1/2$ in (93) is due to the notation $x = |k|^2$. Similarly, for the Laplace transform, we obtain

$$f(v,t) \simeq e^{-\mu(p)t}\Phi_p(ve^{-\mu(p)t}), \qquad t \to \infty, \ ve^{-\mu(p)t} = \text{const.}, \tag{96}$$

where

$$\psi_p(x) = \mathcal{L}[\Phi_p]. \tag{97}$$

The positivity and some other properties of $F_p(|v|)$ follow from the fact that $\psi_p(x) = w_p(x^p)$, where $w_p(x)$ satisfies Theorem 5. Hence

$$\psi_p(x) = \int_0^\infty d\tau\, R_p(\tau)e^{-\tau x^p}, \qquad \int_0^\infty d\tau\, R_p(\tau) = \int_0^\infty d\tau\, R_p(\tau)\tau = 1, \tag{98}$$

where $R_p(\tau)$, $\tau \geq 0$, is a non-negative generalized function (of course, both ψ_p and R_p depend on p).

In particular we can conclude that the self-similar asymptotics (93) for any initial data $f_0 \geq 0$ occurs if $p_0 > 1$, otherwise it occurs for $p \in (0, p_0) \subset (0,1)$. Therefore, for any spectral function $\mu(p)$ (Fig. 1), the approximate equality (93) holds for sufficiently small $0 < p \leq 1$. In addition,

$$m_2 = \int_{\mathbb{R}^d} dv\, f_0(|v|)|v|^2 < \infty \ \text{ if } \ p = 1$$

and $m_2 = \infty$ if $p < 1$. Similar conclusions can be made for the Laplace transforms.

The positivity of $F(|v|)$ in (94)–(97) follows from the integral representation (98) with $p \leq 1$, since it is well-known that

$$\mathcal{F}^{-1}(e^{-|k|^{2p}}) > 0, \qquad \mathcal{L}^{-1}(e^{-x^{2p}}) > 0$$

for any $0 < p \leq 1$ (the so-called infinitely divisible distributions [18]). Thus, (98) explains the connection of the self-similar solutions of generalized Maxwell models with infinitely divisible distributions.

Using standard formulas for the inverse Fourier (Laplace) transforms, denote by ($d = 1, 2, \ldots$ is fixed)

$$M_p(|v|) = \frac{1}{(2\pi)^d} \int_{\mathbb{R}^d} dk\, e^{-|k|^{2p}+ik\cdot v},$$

$$N_p(v) = \frac{1}{2\pi i} \int_{a-i\infty}^{a+i\infty} dx\, e^{-x^p+xv}, \qquad 0 < p \leq 1, \tag{99}$$

we obtain the self-similar solutions (distribution functions) are given in (96), (98) (right-hand sides), by

$$F_p(|v|) - \int_0^\infty d\tau \, R_p(\tau)\tau^{-\frac{d}{2p}} M_p(|v|\tau^{-\frac{1}{2p}}),$$

$$\Phi_p(v) = \int_0^\infty d\tau \, R_p(\tau)\tau^{-\frac{1}{p}} N_p(v\tau^{-\frac{1}{p}}), \qquad v \geq 0, \; 0 < p \leq 1. \tag{100}$$

Note that $M_1(|v|)$ is the standard Maxwellian in $\mathbb{R}^d$. The functions $N_p(v)$ (99) are studied in detail in the literature [18, 20]. Thus, for given $0 < p \leq 1$, the kernel $R(\tau)$, $\tau \geq 0$, is the only unknown function that is needed to describe the distribution functions $F(|v|)$ and $\Phi(v)$. See [10] for a study of $R(\tau)$ in more detail.

Now, from (34), (42), and recalling $\mu = \mu(1)$, we can show the moments equation can be written in the form

$$(s\mu(1) - \lambda(s) + 1)m_s = \sum_{n=2}^{N} \alpha_n I_n(s), \tag{101}$$

where

$$I_n(s) = \int_{\mathbb{R}_+^n} da_1 \ldots da_n A(a_1, \ldots, a_n) \int_{\mathbb{R}_+^n} d\tau_1 \ldots, d\tau_n \, g_n^{(s)}(a_1\tau_1, \ldots, a_n\tau_n) \prod_{j=1}^{n} R(\tau_j)$$

$$\tag{102}$$

$$g_n^{(s)}(y_1, \ldots, y_n) = \left(\sum_{k=1}^{n} y_k\right)^s - \sum_{k=1}^{n} y_k^s, \qquad n = 1, 2, \ldots.$$

and, due to the properties of $R(\tau)$, one gets $g_1^{(s)} = 0$ for any $s \geq 0$ and $m_0 = m_1 = 1$.

Our aim is to study the moments m_s defined in (92), for $s > 1$, on the basis of (101). The approach is similar to the one used in [15] for a simplified version of (101) with $N = 2$. The main results are formulated below in terms of the spectral function $\mu(p)$ (see Fig. 1) under assumption that $p_0 > 1$.

Proposition 2.

(i) If the equation $\mu(s) = \mu(1)$ has the only solution $s = 1$, then $m_s < \infty$ for any $s > 0$.

(ii) If this equation has two solutions $s = 1$ and $s = s_* > 1$, then $m_s < \infty$ for $s < s_*$ and $m_s = \infty$ for $s > s_*$.

(iii) $m_{s_*} < \infty$ only if $I_n(s_*) = 0$ in (101) for all $n = 2, \ldots, N$.

Now we can draw some conclusions concerning the moments of the distribution functions (100) as follows. Denote by

$$m_s(\Phi_p) = \int_0^\infty dv\, \Phi_p(v) v^s, \qquad m_s(R_p) = \int_0^\infty d\tau\, R_p(\tau) \tau^s,$$

$$m_{2s}(F_p) = \int_{\mathbb{R}^d} dv\, F_p(|v|)|v|^{2s}, \qquad s > 0;\; 0 < p \le 1,$$

and use similar notations for $N_p(v)$ and $M_p(|v|)$ in (100). Then, by formal integration of (100), we obtain

$$m_s(\Phi_p) = m_s(N_p) m_{s/p}(R_p)$$

$$m_{2s}(F_p) = m_{2s}(M_p) m_{s/p}(R_p),$$

where M_p and N_p are given in (99) and we show that finite only for $s < p < 1$.

In the remaining case $p = 1$ *all* moments of functions

$$M_1(|v|) = (4\pi)^{-d/2} \exp\left[-\frac{|v|^2}{4}\right], \qquad v \in \mathbb{R}^d;$$

$$N_1(v) = \delta(v - 1), \qquad v \in \mathbb{R}_+,$$

are finite. Therefore, everything depends on moments of R_1 in with $p = 1$, so it only needs to apply Proposition 2.

In particular, the following statement holds for the moments of the distribution functions (93), (96).

Proposition 3.

(i) If $0 < p < 1$, then $m_{2s}(F_p)$ and $m_s(\Phi_p)$ are finite if and only if $0 < s < p$.
(ii) If $p = 1$, then Proposition 2 holds for $m_s = m_{2s}(F_1)$ and for $m_s = m_s(\Phi_1)$.

Remark 8. Proposition 3 can be interpreted in other words: the distribution functions $F_p(|v|)$ and $\Phi_p(v)$, $0 < p \le 1$, can have finite moments of all orders in the only case when two conditions are satisfied

(1) $p = 1$, and
(2) The equation $\mu(s) = \mu(1)$ (see Fig. 1) has the unique solution $s = 1$.

In all other cases, the maximal order s of finite moments $m_{2s}(F_p)$ and $m_s(\Phi_p)$ is bounded.

This fact means that the distribution functions F_p and Φ_p have power-like tails.

6.3 Applications to the Conservative or Dissipative Boltzmann Equation

We recall the three specific Maxwell models (A), (B), (C) of the Boltzmann equation from Sect. 2. Our goal in this section is to study isotropic solutions $f(|v|, t), v \in \mathbb{R}^d$, of (2), (4), and (5) respectively. All three cases are considered below from a unified point of view. First we perform the Fourier transform and denote

$$u(x, t) = \mathcal{F}[f(|v|, t)] = \int_{\mathbb{R}^d} dv\, f(|v|, t) e^{-ik \cdot v}, \quad x = |k|^2, \quad u(0, t) = 1. \quad (103)$$

It was already said at the beginning of Sect. 4 that $u(x, t)$ satisfies (in all three cases) (23), where $N = 2$ and all notations are given in (27), (14), (18)–(21). Hence, all results of our general theory are applicable to these specific models. In all three cases (A), (B), (C) we assume that the initial distribution function

$$f(|v|, 0) = f_0(|v|) \geq 0, \qquad \int_{\mathbb{R}^d} dv\, f_0(|v|) = 1, \quad (104)$$

and the corresponding characteristic function

$$u(0, t) = u_0(x) = \mathcal{F}[f_0(|v|)], \quad x = |k|^2, \quad (105)$$

are given. Moreover, let $u_0(x)$ be such that

$$u_0(x) = 1 - \alpha x^p + O(x^{p+\varepsilon}), \quad x \to 0, \quad 0 < p \leq 1, \quad (106)$$

with some $\alpha > 0$ and $\varepsilon > 0$. We distinguish below the initial data with finite energy (second moment)

$$E_0 = \int_{\mathbb{R}^d} dv\, |v|^2 f_0(|v|) < \infty \quad (107)$$

implies $p = 1$ in (106) and the in-data with infinite energy $E_0 = \infty$. If $p < 1$ in (106) then

$$m_q^{(0)} = \int_{\mathbb{R}^d} dv\, f_0(|v|) |v|^{2q} < \infty \quad (108)$$

only for $q \leq p < 1$ (see [18, 20]).

Also, the case $p > 1$ in (106) is not possible for $f_0(|v|) \geq 0$.

In addition, note that the coefficient $\alpha > 0$ in (106) can always be changed to $\alpha = 1$ by the scaling transformation $\tilde{x} = \alpha^{1/p} x$. Then, without loss of generality, we set $\alpha = 1$ in (106).

Since it is known that the operator $\Gamma(u)$ in all three cases belongs to the class (24), we can apply Theorem 5 and state that self-similar solutions of (23) are given by

$$u_s(x, t) = \Psi(x\, e^{\mu(p)t}), \quad \Psi(x) = w(x^p), \quad (109)$$

where $w(x)$ is given in Theorem 5 and $0 < p < p_0$ (the spectral function $\mu(p)$, defined in (43), and its critical point p_0 depends on the specific model.)

According to Sects. 6.1–6.2, we just need to find the spectral function $\mu(p)$. In order to do this we first define the linearized operator $L = \Gamma'(1)$ for $\Gamma(u)$ given in (24), (27). One should be careful at this point since $A_2(a_1, a_2)$ in (27) is not symmetric and therefore (32) cannot be used. A straight-forward computation leads to

$$Lu(x) = \int_0^1 ds\, G(s)(u(a(s)x) + u(b(s)x)) + \int_0^1 ds H(s) u(c(s)x), \qquad (110)$$

in the notation (18)–(21). Then, the eigenvalue $\lambda(p)$ is given by

$Lx^p = \lambda(p)\, x^p$ which implies

$$\lambda(p) = \int_0^1 ds\, G(s)\left\{(a(s))^p + (b(s))^p\right\} + \int_0^1 ds H(s)(c(s))^p, \qquad (111)$$

and the spectral function (43) reads

$$\mu(p) = \frac{\lambda(p) - 1}{p}. \qquad (112)$$

Note that the normalization (22) is assumed.

At that point we consider the three models (A), (B), (C) separately and apply (111) and (112) to each case.

(A) Elastic Boltzmann Equation (2) in $\mathbb{R}^d, d \geq 2$. By using (18), (19), and (22) we obtain

$$\lambda(p) = \int_0^1 ds\, G(s)(s^p + (1 - s)^p), \qquad G(s) = A_d\, g(1 - 2s)[s(1 - s)]^{\frac{d-3}{2}}, \qquad (113)$$

where the normalization constant A_d is such that (22) is satisfied with $H = 0$. Then

$$\mu(p) = \frac{1}{p} \int_0^1 ds\, G(s)(s^p + (1 - s)^p - 1), \quad p > 0. \qquad (114)$$

It is easy to verify that $p\,\mu(p) \to 1$ as $p \to 0$, $\mu(p) \to 0$ as $p \to \infty$, and

$$\mu(p) > 0 \ \text{ if } p < 1; \qquad \mu(p) < 0 \ \text{ if } p > 1;$$
$$\mu(1) = 0, \qquad \mu(2) = \mu(3) = -\int_0^1 ds\, G(s)\, s(1 - s). \qquad (115)$$

Hence, $\mu(p)$ in this case is similar to the function shown on Fig. 1b with $2 < p_0 < 3$ and such that $\mu(1) = 0$. Then the self-similar asymptotics hold all $0 < p < 1$.

(B) Elastic Boltzmann Equation in the presence of a thermostat (4) in $\mathbb{R}^d$, $d \geq 2$. We consider just the case of a cold thermostat with $T = 0$ in (14), since the general case $T > 0$ can be considered after that with the help of (11). Again, by using (18), (19), and (22) we obtain

$$\lambda(p) = \int_0^1 ds\, G(s)(s^p + (1 - s)^p) + \theta \int_0^1 ds\, G(s)(1 - \frac{4m}{(1 + m)^2})^p,$$

$$G(s) = \frac{1}{1 + \theta} A_d\, g(1 - 2s)[s(1 - s)]^{\frac{d-3}{2}}, \tag{116}$$

with the same constant A_d as in (113). Then

$$\mu(p) = \frac{1}{p} \int_0^1 ds\, G(s)(\, s^p + (1 - s)^p - \theta(1 - \beta s)^p - (1 + \theta)),$$

$$\beta = \frac{4m}{(1 + m)^2}, \quad p > 0, \tag{117}$$

and therefore, as in the previous case (A), $p\,\mu(p) \to 1$ as $p \to 0$, $\mu(p) \to 0$ as $p \to \infty$, with

$$\mu(1) = -\theta\,\beta \int_0^1 ds\, G(s)\, s. \tag{118}$$

which again verifies that $\mu(p)$ is of the same kind as in the elastic case (A) and shown on Fig. 1b. A position of the critical point p_0 such that $\mu'(p_0) = 0$ (see Fig. 1b) depends on θ. It is important to distinguish two cases: (1) $p_0 > 1$ and (2) $p_0 < 1$. In case (1) any non-negative initial data (104) has the self-similar asymptotics. In case (2) such asymptotics holds just for in-data with infinity energy satisfying (106) with some $p < p_0 < 1$. A simple criterion to separate the two cases follows directly from Fig. 1b: it is enough to check the sign of $\mu'(1)$. If

$$\mu'(1) = \lambda'(1) - \lambda(1) + 1 < 0 \tag{119}$$

in the notation of (116), then $p_0 > 1$ and the self-similar asymptotics hold for any non-negative initial data.

The inequality (119) is equivalent to the following condition on the positive coupling constant θ

$$0 < \theta < \theta_* = -\frac{\int_0^1 ds\, G(s)\, (s \log s + (1 - s) \log(1 - s))}{\int_0^1 ds\, G(s)\, (\beta s + (1 - \beta s) \log(1 - \beta s))}. \tag{120}$$

The right-hand side of this inequality is positive and independent on the normalization of $G(s)$, therefore it does not depend on θ (see (117). We note that a new class of exact self-similar solutions to (4) with finite energy was recently found in [12] for $\beta = 1, \theta = 4/3$ and $G(s) = const$. A simple calculation of the integrals in (120) shows that $\theta_* = 2$ in that case, therefore the criterion (119) is fulfilled for the exact solutions from [12] and they are asymptotic for a wide class of initial data with finite energy. Similar conclusions can be made in the same way about exact positive self-similar solutions with infinite energy constructed in [12]. Note that the inequality (119) shows the non-linear character of the self-similar asymptotics: it holds unless the linear term in (4) is "too large".

(C) Inelastic Boltzmann Equation (5) in $\mathbb{R}^d$. Equations (21) and (22) lead to

$$\lambda(p) = \int_0^1 ds \, G(s)((a\,s)^p + (1 - b\,s)^p),$$

where

$$G(s) = C_d \,(1 - s)^{\frac{d-3}{2}}, \qquad a = \frac{(1 + e)^2}{4}, \qquad b = \frac{(1 + e)(3 - e)}{4},$$

with such constant C_d that (22) with $H = 0$ is fulfilled. Hence

$$\mu(p) = \frac{1}{p} \int_0^1 ds \, G(s)((a\,s)^p + (1 - b\,s)^p - 1), \qquad p > 0, \qquad (121)$$

and once more as in the previous two cases, $p\,\mu(p) \to 1$ as $p \to 0$, $\mu(p) \to 0$ as $p \to \infty$, with now

$$\mu(1) = -\frac{1 - e^2}{4} \int_0^1 ds \, G(s) \, s.$$

Thus, the same considerations lead to the shape of $\mu(p)$ shown in Fig. 1b. The inequality (119) with $\lambda(p)$ given in (121) was proved in [11] (see (4.26) of [11], where the notation is slightly different from ours). Hence, the inelastic Boltzmann equation (5) has self-similar asymptotics for any restitution coefficient $0 < e < 1$ and any non-negative initial data.

Hence, the spectral function $\mu(p)$ in all three cases above is such that $p_0 > 1$ provided the inequality (120) holds for the model (B).

Therefore, according to our general theory, all "physical" initial conditions (104) satisfying (106) with any $0 < p \leq 1$ lead to self-similar asymptotics. Hence, the main properties of the solutions $f(v, t)$ are qualitatively similar for all three models (A), (B) and (C), and can be described in one unified statement: Theorem 6 below.

Before we formulate such general statement, it is worth to clarify one point related to a special value $0 < p_1 \leq 1$ such that $\mu(p_1) = 0$. The reader can see that on Fig. 1b that the unique root of this equation exits for all models (A), (B), (C) since $\mu(1) = 0$ in the case (A) (energy conservation), and $\mu(1) < 0$ in cases (B) and (C) (energy dissipation). If $p = p_1$ in (106) then the self-similar solution (109) is simply a stationary solution of (23). Thus, the time relaxation to the non-trivial ($u \neq 0, 1$) stationary solution is automatically included in Theorem 6 as a particular case of self-similar asymptotics.

Thus we consider simultaneously (2), (4), (5), with the initial condition (104) such that (106) is satisfied with some $0 < p \leq 1$, $\varepsilon > 0$ and $\alpha = 1$. We also assume that $T = 0$ in (4) and the coupling parameter $\theta > 0$ satisfies the condition (120).

In the following Theorem 6, the solution $f(|v|, t)$ is understood in each case as a generalized density of probability measure in $\mathbb{R}^d$ and the convergence $f_n \to f$ in the sense of weak convergence of probability measures.

Theorem 6. *The following two statements hold*

(i) *There exists a unique (in the class of probability measures) solution $f(|v|, t)$ to each of (2), (4), (5) satisfying the initial condition (104). The solution $f(|v|, t)$ has self-similar asymptotics in the following sense: For any given $0 < p \leq 1$ in (106) there exits a unique non-negative self-similar solution*

$$f_s^{(p)}(|v|, t) = e^{-\frac{d}{2}\mu(p)\,t} F_p(|v| e^{-\frac{1}{2}\mu(p)\,t}), \tag{122}$$

such that

$$e^{\frac{d}{2}\mu(p)\,t} f(|v| e^{-\frac{1}{2}\mu(p)\,t}, t) \to_{t\to\infty} F_p(|v|), \tag{123}$$

where $\mu(p)$ is given in (114), (117), (121), respectively, for each of the three models.

(ii) *Except for the special case of the Maxwellian*

$$F_1(|v|) = M(|v|) = (4\pi)^{-d/2} e^{-\frac{|v|^2}{4}} \tag{124}$$

for (2) with $p = 1$ in (106) (note that $\mu(1) = 0$ in this case), the function $F_p(|v|)$ does not have finite moments of all orders. If $0 < p < 1$, then

$$m_q = \int_{\mathbb{R}^d} dv\, F_p(|v|)|v|^{2q} < \infty \qquad \text{only for } 0 < q < p. \tag{125}$$

If $p = 1$ in the case of (4), (5), then $m_q < \infty$ only for $0 < q < p_*$, where $p_* > 1$ is the unique maximal root of the equation $\mu(p_*) = \mu(1)$, with $\mu(p)$ given in (106), (121) respectively.

In addition, we also we obtain the following corollary.

Corollary 6. *Under the same conditions of Theorem 6, the following two statements hold.*

(i) *The rate of convergence in (123) is characterized in terms of the corresponding characteristic functions in Proposition 1.*

(ii) *The function $F_p(|v|)$ admits the integral representation (100) through infinitely divisible distributions (99).*

Proof. It is enough to note that all results of Sects. 6.1 and 6.2 are valid, in particular, for (2), (4), (5).

Finally, we mention that the statement similar to Theorem 6, can be easily derived from general results of Sect. 6.1 in the case of one-dimensional Maxwell models introduced in [2,15] for applications to economy models (Pareto tails, etc.). The only difference is that the "kinetic" equation can be transformed to its canonical form (23)–(24) by the Laplace transform and that the spectral function $\mu(p)$ can have in this case any of the four kind of behaviors shown in Fig. 1. The only remaining problem for any such one-dimensional models is to study them for their specific function $\mu(p)$, and then to apply Propositions 1, 2 and 3.

Thus, the general theory developed in this chapter is applicable to all existing multi-dimensional isotropic Maxwell models and to one-dimensional models as well.

Acknowledgement

The first author was supported by grant 621-2003-5357 from the Swedish Research Council (NFR). The research of the second author was supported by MIUR of Italy. The third author has been partially supported by NSF under grant DMS-0507038. Support from the Institute for Computational Engineering and Sciences at the University of Texas at Austin is also gratefully acknowledged.

References

1. A. Baldassarri, U.M.B. Marconi, and A. Puglisi; Influence of correlations on the velocity statistics of scalar granular gases, Europhys. Lett., 58 (2002), pp. 14–20.
2. Ben-Abraham D., Ben-Naim E., Lindenberg K., Rosas A.; Self-similarity in random collision processes, Phys. Review E, 68, R050103 (2003).
3. Ben-Naim E., Krapivski P.; Multiscaling in inelastic collisions, Phys. Rev. E, R5–R8 (2000).
4. Bisi M., Carrillo J.A., Toscani G.; Decay rates in probability metrics towards homogeneous cooling states for the inelastic Maxwell model, *J. Stat. Phys.* 118 (2005), no. 1–2, 301–331.
5. A.V. Bobylev; The theory of the nonlinear spatially uniform Boltzmann equation for Maxwell molecules. Mathematical Physics Reviews, Vol. 7, 111–233, Soviet Sci. Rev. Sect. C Math. Phys. Rev., 7, Harwood Academic Publ., Chur, (1988).
6. A.V. Bobylev; The Fourier transform method for the Boltzmann equation for Maxwell molecules, Sov. Phys. Dokl. 20: 820–822 (1976).
7. A.V. Bobylev, J.A. Carrillo, and I.M. Gamba; On some properties of kinetic and hydrodynamic equations for inelastic interactions. J. Statist. Phys. 98 (2000), no. 3–4, 743–773.
8. A.V. Bobylev and C. Cercignani; Self-similar solutions of the Boltzmann equation and their applications, J. Statist. Phys. 106 (2002), 1039–1071.

9. A.V. Bobylev and C. Cercignani; Self-similar asymptotics for the Boltzmann equation with inelastic and elastic interactions, J. Statist. Phys. 110 (2003), 333–375.

10. A.V. Bobylev, C. Cercignani, and I.M. Gamba; On the self-similar asymptotics for generalized non-linear kinetic Maxwell models, To appear in Comm. Math. Physics. http://arxiv.org/abs/math-ph/0608035 (2006).

11. A.V. Bobylev, C. Cercignani, and G. Toscani; Proof of an asymptotic property of self-similar solutions of the Boltzmann equation for granular materials, J. Statist. Phys. 111 (2003), 403–417.

12. A.V. Bobylev and I.M. Gamba; Boltzmann equations for mixtures of Maxwell gases: exact solutions and power like tails, *J. Stat. Phys.* 124, no. 2–4, 497–516. (2006).

13. N. Brilliantov and T. Poschel (Eds.); Granular Gas Dynamics, Springer, Berlin, 2003.

14. C. Cercignani; The theory of the nonlinear spatially uniform Boltzmann equation for Maxwell molecules, *Sov. Sci. Rev. C. Math. Phys.*, **7**, 111–233 (1988).

15. S. Cordier, L. Pareschi, and G. Toscani; On a kinetic model for a simple market economy. J. Statist. Phys. 120 (2005), 253–277.

16. M.H. Ernst and R.Brito; Scaling solutions of inelastic Boltzmann equations with overpopulated high energy tails; J. Stat. Phys. 109 (2002), 407–432.

17. M. Ernst, E. Trizac, and A. Barrat, Granular gases: Dynamics and collective effects, Journal of Physics C: Condensed Matter. condmat-0411435. (2004).

18. W. Feller; An Introduction to Probability Theory and Applications, Vol. 2, Wiley, N.-Y., 1971.

19. E. Gabetta, G. Toscani, and W. Wennberg; Metrics for probability distributions and the trend to equilibrium for solutions of the Boltzmann equation, J. Stat. Phys. 81:901 (1995).

20. E. Lukacs; Characteristic Functions, Griffin, London 1970.

21. G. Menon and R. Pego; Approach to self-similarity in Smoluchowski's coagulation equations. Comm. Pure Appl. Math. 57 (2004), no. 9, 1197–1232.

22. G. Menon and R. Pego; Dynamical scaling in Smoluchowski's coagulation equations: uniform convergence, no. 5, 1629–1651, SIAM Applied Math, 2005.

Hydrodynamics from the Dissipative Boltzmann Equation

Giuseppe Toscani

Summary. In this chapter we discuss various questions related with the modeling of hydrodynamic equations for granular gases, starting from the kinetic description based on the dissipative Boltzmann equation. A comparison with the elastic case is briefly presented, together with the main open problems.

1 Introduction

This chapter deals with some questions related with the modeling of hydrodynamic equations for granular gases, at the light of recent mathematical results on the large-time behavior of the dissipative Boltzmann equation. This subject is relatively new, and the relevant mathematical theory is still restricted. In the pertinent literature [Duf01], rapid granular flows were frequently described at the macroscopic level by means of equations for fluid dynamics, modified to account for dissipation due to collisions among particles. This was the approach of Haff, which, in his pioneering paper [Haf83], gave a macroscopic description of the behavior of a granular material treating the individual grains as the molecules of a granular fluid, without resorting to the mesoscopic picture (the Boltzmann or Enskog kinetic equations).

In more recent years it became clear that, in agreement with the well established derivation of conservative fluid dynamics from the Boltzmann equation [BGL91, BGL93], kinetic theory was the basis for a deeper understanding of macroscopic equations even for dissipative flows. Kinetic theory is suitable to describe the evolution of materials composed of many small discrete grains, in which the mean free path of the grains is much larger than the typical particle size. In this regime, granular gases can be described within the concepts of classical statistical mechanics, by adapting methods of the kinetic theory of ideal gases [Kog69, CIP94].

Many authors (see [NY93, DLK95, BCP97, EP97, BCG00, BP00a] and the references therein) adopted this line of thought, by introducing and discussing the evolution of a system of partially inelastic rigid spheres through Boltzmann-like equations. A typical kinetic model for the study of the evolution of a granular material takes the following form: the unknown is a time dependent density in phase space $f(x, v, t)$ satisfying a Boltzmann–Enskog equation for inelastic hard-spheres, which for the force-free case reads

$$\frac{\partial f}{\partial t} + v \cdot \nabla_x f = \frac{1}{\epsilon} C(f, f)(x, v, t). \tag{1}$$

Here $v \cdot \nabla_x f$ is the usual transport operator where $C(f, f)$ is the so-called granular collision operator, which describes the change in the density function due to creation and annihilation of particles in dissipative binary collisions. The ϵ-parameter (Knudsen number) represents a measure of the mean free path, and has to be assumed small in fluid dynamical regimes.

The loss of energy in the microscopic collision translates at a macroscopic level in the progressive cooling of the gas, a phenomenon which is responsible of most of the difficulties in extending methods of classical kinetic theory of ideal gases to granular ones. A clear understanding of the new problems one has to deal with in the derivation of macroscopic equations in dissipative kinetic theory can by obtained through the use of the splitting method, very popular in the numerical approach to the Boltzmann equation [GPT97,PR01]. If at each time step we consider sequentially the transport and relaxation operators in the Boltzmann equation (1), during this short time interval we recover the evolution of the density from the joint action of the relaxation

$$\frac{\partial f}{\partial t} = \frac{1}{\epsilon} C(f, f)(x, v, t), \tag{2}$$

and transport

$$\frac{\partial f}{\partial t} + v \cdot \nabla_x f = 0. \tag{3}$$

In classical kinetic theory, the energy is conserved in collisions, and the relaxation (2) pushes the solution towards the Maxwellian equilibrium with the same mass, momentum and energy of the initial datum. Then, if ϵ is sufficiently small, one can easily argue that the solution to (2) is *sufficiently close* to the Maxwellian, and this Maxwellian can be used into the transport step (3). When dissipation is present, solutions of the inelastic Boltzmann equation lose energy until all particles travel at the same speed, and the relaxation (2) pushes the solution towards the asymptotic state represented by a δ function concentrated in the mean velocity of the initial value. It is evident that, if $\epsilon \ll 1$, so that the solution to (2) is close to this *poor* asymptotic state, substitution into the transport step (3) does not lead to any correct behavior.

To circumvent this difficulty, two different procedures have been proposed. The first one requires that microscopic collisions are weakly inelastic. In this case, one assumes that the collision operator $C(f, f)$ can be decomposed as

$$C(f, f) = B(f, f) + \beta I(f, f), \tag{4}$$

where $B(f, f)$ is the elastic collision operator, while $I(f, f)$ represents the inelastic correction. If β is of the same order of ϵ, so that $\beta/\epsilon \to \lambda$ as $\beta, \epsilon \to 0$, we can easily modify in this case the aforementioned splitting, putting the elastic collision operator in the relaxation step

$$\frac{\partial f}{\partial t} = \frac{1}{\epsilon} \, B(f,f)(x,v,t), \tag{5}$$

and including the granular correction into the transport

$$\frac{\partial f}{\partial t} + v \cdot \nabla_x f = \frac{\beta}{\epsilon} \, I(f,f). \tag{6}$$

As before, the relaxation (5) pushes the solution towards the Maxwellian equilibrium with the same mass, momentum and energy of the initial datum. Then, if ϵ is sufficiently small, $\beta/\epsilon \cong \lambda$, and the Maxwellian solution can be used in the transport step (6), to get explicitly computable equations for the macroscopic quantities.

A second method is based on a more precise study of the asymptotic behavior of the solution in the relaxation step. This is obtained by looking for exact solutions to (2) (homogeneous cooling states), with the aim to use this exact solution in the transport (3).

The main advantage in working with small inelasticity is that one can easily include in the procedure any type of inelastic collisions, and in particular general coefficients of restitution. Various studies enlighten in fact the dependence of the cooling problem on the coefficient of restitution in the microscopic collision, and emphasize the effects of a non-constant restitution coefficient [BP00a, BP03a]. Special attention has been devoted in this respect to a system of viscoelastic spheres, a quite realistic model whose coefficient of restitution has been recently derived [RPBS99]. Moreover, the elastic collision dominated regime prevents the derivation to be sensible to the strength of spatial gradients. From a mathematical point of view, the granular Boltzmann equation has been object of some attention in a recent past [BC02b, BC03], and hydrodynamic closure in a weakly inelastic regime has been discussed already in [GS95, BCG00, BDKS98, Tos04], mostly at the level of Euler equations. A further step beyond the Euler level, in which closure is achieved by simply using the zero order solution (the equilibrium Maxwellian) for the distribution function, has been recently done in [RC02, BST04], by resorting to a Grad 13-moment expansion able to capture the relevant Grad equations. The idea of applying Grad's method to inelastic gases goes back to Jenkins and Richmann [JR85]. In this pioneering paper they outline the main ideas of Grad's derivation of hydrodynamics from a kinetic equation, using the Maxwellian distribution to close the hierarchy of transport equations. An important feature of Grad's equations is that they still contain collision terms, and are affected by the same small parameters as the kinetic equations. They lend themselves then to a classical asymptotic procedure of the Chapman–Enskog type, and provide as important byproduct hydrodynamic equations at the Navier–Stokes level. We refer to the recent paper [GK02] for a detailed treatment of the classical Chapman–Enskog derivation of hydrodynamics given in the framework of Grad's moment equations.

As far as the second method is concerned, the common assumption which has been at the basis of several recent papers on the matter is that there are only small spatial variations, so that the zero order approximation of the solution (and of any asymptotic expansion) is constituted by the so-called homogeneous cooling state (see for instance [BDKS98] and the references therein). A detailed theory of the homogeneous cooling state for viscoelastic particles in terms of expansion in Sonine polynomials has been recently developed [BP00b, BP00c]. Such spatially homogeneous solution turns out to depend not only on the similarity variable, as it would occur for constant restitution coefficient, but also on time explicitly. In addition, temperature has been shown to decay asymptotically according to a corrected Haff's law (see also [SP98]). Asymptotic expansions around the homogeneous cooling state have been used then as hydrodynamic closure for the macroscopic equations in order to achieve a Navier–Stokes level via a Chapman–Enskog procedure also for non-constant restitution coefficient [LS86]. In particular, the complete set of hydrodynamic equations and transport coefficients have been derived in this frame for a granular gas of viscoelastic particles [BP02, BP03b]. From a mathematical point of view, the possibility of using the homogeneous cooling state to close the transport equation (3) would require precise statements on the role of this exact solution, which has to be the intermediate solution of the homogeneous problem for a large *physical* class of initial densities. For a simplified model of the Boltzmann equation, the so-called Maxwellian model [BCG00], recent results showed that this holds true [BCT03, BCT05a]. These results were first motivated by a question posed by Ernst and Brito [EB02a, EB02b], concerning the fat tails of the self-similar solution for this model.

2 Modeling Dissipative Boltzmann Equation

In a granular gas, the microscopic dynamics of grains is governed by the restitution coefficient e which relates the normal components of the particle velocities before and after a collision. If grains are identical perfect spheres of diameter $\sigma > 0$, (x, v) and $(x - \sigma n, w)$ are their states before a collision, where $n \in S^2$ is the unit vector along the center of both spheres, the post collisional velocities (v^*, w^*) are such that

$$(v^* - w^*) \cdot n = -e((v - w) \cdot n). \tag{7}$$

Thanks to (7), and assuming the conservation of momentum, one finds the change of velocity for the colliding particles as

$$v^* = v - \frac{1}{2}(1 + e)((v - w) \cdot n)n, \quad w^* = w + \frac{1}{2}(1 + e)((v - w) \cdot n)n. \tag{8}$$

For elastic collisions one has $e = 1$, while for inelastic collisions e decreases with increasing degree of inelasticity.

In the literature, it is frequently assumed that the restitution coefficient is a physical constant. In real applications, however, the restitution coefficient may depend on the relative velocity in such a way that collisions with small relative velocity are close to be elastic. The simplest physically correct description of dissipative collisions is based on the assumption that the spheres are composed by viscoelastic material [BP00a, BP00d]. This analysis suggests that in general the coefficient of restitution is such that

$$1 - e = 2\beta\gamma\left(|(v - w) \cdot n|\right), \tag{9}$$

where $\gamma(\cdot)$ is a given function and β is a parameter which is small in presence of small inelasticity. For example, for small values of α, the velocity dependence of the restitution coefficient in a collision of viscoelastic spheres can be expressed at the leading order as in (9), choosing $\gamma(r) = r^{1/5}$ [BP00a]. In a rarefied regime, a general model of bilinear operator for dissipative collisions is obtained by choosing

$$C(f, f)(x, v, t) = G(\rho)\bar{Q}(f, f)(x, v, t), \tag{10}$$

where $\bar{Q}$

$$Q(f, f)(v) = 4\sigma^2 \int_{I\!R^3} \int_{S_+} q \cdot n \left\{\chi f(v^{**})f(w^{**}) - f(v)f(w)\right\} dw\, dn. \tag{11}$$

In (10)

$$\rho(x, t) = \int_{I\!R^3} f(x, v, t)\, dv \tag{12}$$

is the density, and the function $G(\rho)$ is the statistical correlation function between particles, which accounts for the increasing collision frequency due to the excluded volume effects. We refer to [Cer95] for a detailed discussion of the meaning of the function G.

In (11), $q = (v - w)$, and S_+ is the hemisphere corresponding to $q \cdot n > 0$. The velocities (v^{**}, w^{**}) are the pre collisional velocities of the so-called inverse collision, which results with (v, w) as post collisional velocities. The factor χ in the gain term appears respectively from the Jacobian of the transformation $dv^{**} dw^{**}$ into $dv dw$ and from the lengths of the collisional cylinders $e|q^{**} \cdot n| = |q \cdot n|$. For a constant restitution coefficient, $\chi = e^{-2}$.

In what follow we write the operator (11) in weak form [Tos04]. More precisely, for all smooth functions $\varphi(v)$, it holds

$$<\varphi, \bar{Q}(f, f)> = \int_{I\!R^3} \varphi(v)\bar{Q}(f, f)(v)\, dv =$$

$$4\sigma^2 \int_{I\!R^3} \int_{I\!R^3} \int_{S_+} q \cdot n \left(\varphi(v^*) - \varphi(v)\right) f(v)f(w) dv\, dw\, dn =$$

$$2\sigma^2 \int_{I\!R^3} \int_{I\!R^3} \int_{S^2} |q \cdot n| \left(\varphi(v^*) - \varphi(v)\right) f(v)f(w) dv\, dw\, dn. \tag{13}$$

The last equality follows since the integral over the hemisphere S_+ can be extended to the entire sphere S^2, provided the factor $1/2$ is inserted in front of the integral itself. Let (v', w') be the post collisional velocities in a elastic collision with (v, w) as incoming velocities,

$$v' = v - (q \cdot n)n, \quad w' = w + (q \cdot n)n. \tag{14}$$

Using (8) and (14) one obtains

$$v^* = v' + \frac{1}{2}(1 - e)(q \cdot n)n, \quad w^* = w' - \frac{1}{2}(1 - e)(q \cdot n)n. \tag{15}$$

If we assume that the coefficient of restitution satisfies (9),

$$v^* - v' = \beta\gamma\left(|q \cdot n|\right)(q \cdot n)n. \tag{16}$$

Let us consider a Taylor expansion of $\varphi(v^*)$ around $\varphi(v')$. Thanks to (16) we get

$$\varphi(v^*) = \varphi(v') + \beta\nabla\varphi(v') \cdot \gamma\left(|q \cdot n|\right)(q \cdot n)n + O(\beta). \tag{17}$$

If the collisions are nearly elastic, $\beta \ll 1$, and we can cut the expansion (17) after the first-order term. Inserting (17) into (13) gives

$$< \varphi, \bar{Q}(f, f) > = 2\sigma^2 \int_{{I\!R}^3} \int_{{I\!R}^3} \int_{S^2} |q \cdot n| \times$$
$$(\varphi(v') - \varphi(v) + \beta\nabla\varphi(v') \cdot \gamma\left(|q \cdot n|\right)(q \cdot n)n)\, f(v)f(w)dv\, dw\, dn =$$
$$< \varphi, Q(f, f) > + \beta < \varphi, I(f, f) > . \tag{18}$$

It is a simple matter to recognize that in (18) $Q(f, f)$ is the classical Boltzmann collision operator for elastic hard-spheres molecules [CIP94, CK70],

$$Q(f, f)(v) = 2\sigma^2 \int_{{I\!R}^3} \int_{S^2} |q \cdot n| \left\{f(v')f(w') - f(v)f(w)\right\} dw\, dn. \tag{19}$$

In fact, the velocity v' into (18) is obtained from (v, w) through the elastic collision (14).

The second contribution to the inner product (18) can be easily computed [Tos04]. For weak inelasticity the granular correction is the nonlinear friction operator $\beta I(f, f)(v)$, where

$$I(f, f)(v) = 2\sigma^2 \mathrm{div}_v \int_{{I\!R}^3} \int_{S^2} n(q \cdot n)|q \cdot n|\gamma\left(|q \cdot n|\right) f(v')f(w')dw\, dn. \tag{20}$$

The nonlinear friction operator is such that mass and momentum are collisional invariant, while the energy is not. If the restitution coefficient satisfies (9), the Enskog–Boltzmann equation can be modeled at the leading order as

$$\frac{\partial f}{\partial t} + v \cdot \nabla_x f = G(\rho)Q(f, f)(x, v, t) + G(\rho)\beta I(f, f)(x, v, t), \tag{21}$$

where Q is the classical elastic Boltzmann collision operator, and I is a dissipative nonlinear friction operator which is based on elastic collisions between particles. An interesting property of the collisional integral I, connected with the passage to fluid dynamics, is that it leads to exact computations in correspondence to a locally Maxwellian function

$$M(x, v, t) = \frac{\rho(x, t)}{(2\pi T(x, t))^{3/2}} \exp\left(-\frac{(v - u(x, t))^2}{2T(x, t)}\right). \tag{22}$$

In (22) $\rho(x, t)$, $u(x, t)$ and $T(x, t)$ represent its mass, mean velocity and temperature, respectively. In particular one can explicitly evaluate moments. Let us consider the case in which $\gamma(r) = r^p$, with $p \geq 0$. Then [Tos04]

$$< \frac{1}{2}v^2, I(M, M) > = -\frac{\Gamma(2 + p/2)2^{7+p}\sqrt{\pi}}{(4 + p)}\rho^2 T^{(3+p)/2}. \tag{23}$$

3 Hydrodynamic Limit and the Euler Equations

Let us fix the coefficient of restitutions to satisfy $\gamma(r) = r^p$. This choice will include both the constant coefficient of restitution and the physically relevant case of the viscoelastic spheres. The procedure discussed in the introduction allows to formally derive the fluid dynamical equations in the regime of small inelasticity. Since Q is the classical elastic Boltzmann collision operator, from (2) we obtain that the solution is close to the Maxwellian equilibrium with the same mass, momentum and energy of the initial datum provided ϵ is sufficiently small. Inserting this Maxwellian into the transport step (3), gives

$$\int_{\mathbb{R}^3} \psi(v)\left(\frac{\partial M}{\partial t} + v \cdot \nabla_x M - g(\rho)\frac{\beta}{\epsilon}I(M, M)(x, v, t)\right) dv = 0, \tag{24}$$

for any test function ψ. It is well-known that system (24) for the moments of f, which is in general not closed, is closed by assuming f to be a locally Maxwellian function like (22) [CIP94]. Since the dissipative operator I is such that $\psi = 1, v$ are collisional invariants, choosing $\psi = 1, v, \frac{1}{2}v^2$ we obtain from (24) the dissipative Euler equations for density $\rho(x, t)$, bulk velocity $u(x, t)$ and temperature $T(x, t)$

$$\frac{\partial \rho}{\partial t} + \text{div}(\rho u) = 0$$

$$\frac{\partial u}{\partial t} + (u \cdot \nabla)u + \frac{1}{\rho}\nabla p = 0 \tag{25}$$

$$\frac{\partial T}{\partial t} + (u \cdot \nabla)T + \frac{2}{3}T\text{div}u = -\frac{\beta}{\epsilon}C_p g(\rho)\rho T^{(3+p)/2}$$

where $p = \rho T$, and

$$C_p = \frac{\Gamma(2 + p/2)2^{7+p}\sqrt{\pi}}{3(4 + p)}, \tag{26}$$

This approximation is valid when both $\epsilon \ll 1$, $\beta \ll 1$ in such a way that $\beta/\epsilon = \lambda$. This is clearly a nearly elastic regime. If we assume this relationship between β and ϵ, and $p = 1/5$, we obtain the Euler system for a weakly dissipative system of viscoelastic spheres

$$\frac{\partial \rho}{\partial t} + \mathrm{div}(\rho u) = 0$$

$$\frac{\partial u}{\partial t} + (u \cdot \nabla)u + \frac{1}{\rho}\nabla p = 0 \tag{27}$$

$$\frac{\partial T}{\partial t} + (u \cdot \nabla)T + \frac{2}{3}T\mathrm{div}u = -\lambda C_{1/5}g(\rho)\rho T^{8/5}$$

Further applications of this idea lead to higher order hydrodynamic equations [BST04]. Now, the sought approximate closure for the collision term is achieved by replacing, in the evaluation of higher order moments, both in $Q(f, f)$ and $I(f, f)$ the actual distribution function f with the Grad distribution function [Gra49], which, in the spatially one-dimensional case, reads as

$$f_G(\mathbf{v}) = \frac{\rho}{(2\pi T)^{\frac{3}{2}}} e^{-\frac{c^2}{2T}} \left[1 + \frac{p}{2\rho T^2}\left(-\frac{1}{2}c^2 + \frac{3}{2}c_z^2\right) + \frac{4}{5}\frac{q}{2\rho T^2}c_z\left(\frac{c^2}{2T} - \frac{5}{2}\right) \right], \tag{28}$$

and constitutes the weighted polynomial approximation to f sharing the same moments up to heat flux. Consistently with our hypothesis of small ϵ and small β, (28) represents a perturbation to a Maxwellian distribution, solution to the elastic problem in the hydrodynamic limit.

We remark that at present there are no rigorous results which justify the derivation of macroscopic equations in the case of dissipative collisions. As a matter of fact, in classical elastic kinetic theory, one of the main ingredients in the proof is the well-known Boltzmann H-theorem, which guarantees that, at any fixed positive time, the solution is close to the Maxwellian equilibrium provided the Knudsen number is small enough. In recent years, many efforts have been done to obtain explicit computable formulas which allow to quantify the space homogeneous time decay of the solution towards the Maxwellian in terms of the time decay of the relative entropy

$$H(f|M) = \int_{I\!\!R^3} f(v) \log \frac{f(v)}{M(v)} \, dv, \tag{29}$$

where M is the Maxwellian function with the same constant mass ρ, drift velocity u and temperature T of f. In particular, lower bounds on the entropy production

$$-D(f) = \int_{I\!\!R^3} \log f(v)Q(f, f)(v)dv$$

in terms of the relative entropy have been obtained in [TV99a].

The main problem here is the lack of H-theorem for the dissipative Boltzmann equation. A semi-formal discussion on the behavior of the

Boltzmann entropy in this case can be found in [BST04]. We only recall that, since the granular gas is cooling, the correct relative entropy in this case should be

$$H(f|M)(t) = \int_{I\!R^3} f(v,t) \log f(v,t)\, d\mathbf{v} - \rho \log \rho + \frac{3}{2}\rho \log\Big[2\pi e\, T(t)\Big], \quad (30)$$

with $H(f|M) \geq 0$.

4 Hydrodynamics from Homogeneous Cooling States

As described in the previous section, the macroscopic description of a rapid granular flow in a weakly inelastic regime by means of the closure with a local Maxwellian is easy to handle, and reasonable in many physical situations. The alternative to this approach relies in closing the equations with respect to the homogeneous cooling state, namely an exact solution to the dissipative Boltzmann equation. In principle, this alternative way of closure requires a knowledge of its stability which in many cases is far from being understood.

In many fields of evolution equations, exact solutions of self-similar type play an important role as attractors of wide classes of other solutions. Among others, this is the case of nonlinear diffusions, where Barenblatt type self-similar solutions have been recognized as intermediate solutions for the large-time behavior of porous medium equations [Vaz83, Vaz03]. Unlikely, this is not the case in kinetic theory of rarefied gases, where various examples show that self-similar solutions are not always stable with respect to a reasonably large class of physical data. The first example is furnished by the well-known Bobylev–Krook–Wu mode of the Maxwell–Boltzmann equation [Bob88, KW76], a special self-similar solution of the elastic Boltzmann equation for Maxwellian molecules. In fact, even if has been conjectured that the BKW-mode attracts solutions of the Boltzmann equation for a large class of initial data, before these solutions reach the Maxwellian equilibrium, this conjecture has never been verified. More recently, Caglioti and Villani [CV02] gave a precise mathematical proof of the very weak stability properties of the self-similar solution of the nonlinear friction equation introduced by McNamara and Young [NY93, BCP97] as one-dimensional model for the cooling of a granular gas. After their analysis, it is obvious to questioning about the role of homogeneous cooling states, at least in the case of too simplified dissipative models.

As far as the hydrodynamics closure is concerned, the first step in the validation of the closure around the homogeneous cooling state requires a proof of the intermediate asymptotic property of the cooling state itself. In other words, one has to be sure that, (at least in homogeneous situations) if $f_\infty(v,t)$ denotes the homogeneous cooling state of (2), the solution to this equation is such that

$$f(v, t/\epsilon) = f_\infty(v, t/\epsilon) + O(\epsilon^2). \tag{31}$$

Relation (31) insures that the solution to the dissipative Boltzmann equation reaches the self-similar profile before reaching the asymptotic state which has the form of a Dirac delta. A proof of this result is difficult due to the fact that there is no H-theorem, while this theorem is at the basis of similar results in the nonlinear diffusion case [CT00, CJMTU01].

The program has been realized in the case of solutions of the homogeneous Boltzmann equation for the inelastic Maxwell molecules introduced in [BCG00], where in (1)

$$C(f, f) = B\sqrt{\theta(t)}\tilde{Q}(f, f). \tag{32}$$

Here, $\tilde{Q}(f, f)$ is the inelastic Boltzmann collision operator with Maxwellian molecules,

$$(\varphi, \tilde{Q}(f, f)) = \frac{1}{4\pi} \int_{I\!R^3} \int_{I\!R^3} \int_{S^2} f(v)f(w)\Big[\varphi(v') - \varphi(v)\Big] dv\, dw\, dn. \tag{33}$$

In expression (32), the factors B and the temperature of f in front of Q,

$$\theta(t) = \frac{1}{3} \int_{I\!R^3} |v|^2 f(v, t)\, dv,$$

allow the Maxwell model to have the same loss of temperature law of the inelastic hard-spheres model (11).

In (33) the outgoing velocities assumed by a particle in the collision defined by the ingoing velocities v, w and the angular parameter $n \in S^2$:

$$v' = \frac{1}{2}(v + w) + \frac{1-e}{4}(v - w) + \frac{1+e}{4}|v - w|n, \tag{34}$$

$$w' = \frac{1}{2}(v + w) - \frac{1-e}{4}(v - w) - \frac{1+e}{4}|v - w|n. \tag{35}$$

The restitution coefficient e is here assumed to be constant.

Inelastic Maxwell models are of interest for granular fluids in spatially homogeneous states because of the mathematical simplifications resulting from their energy-independent collision rate. For this reason, after its introduction in [BCG00], (32) has been widely studied with or without energy supplies.

Among others, one of the interesting features of granular flows, which can be observed in the framework of Maxwellian molecules, is the knowledge of many of the properties of self-similar solutions in the homogeneous cooling problem, and the non-Maxwellian behavior of these solutions, which display power-like decay for large velocities. Inelastic Maxwell models allow one to take advantage of the powerful Fourier transform methods. Using these techniques for the self-similar scaling problem, existence of solutions with power-like tails has been proven by several authors [BMP02, BK00, BK02, EB02a]. A systematic approach to the existence of self-similar profiles for both elastic and inelastic interactions was subsequently proposed by Bobylev and Cercignani in [BC03], who obtained also results of convergence towards the self-similar solution. Later on, these results have been improved in [BCT03], by showing that convergence towards the self-similar profile occurs for all solutions corresponding to initial data which have more that two moments bounded. In both papers, however, no rate of convergence was found. We remark here that, in view of the passage to hydrodynamics, the knowledge of the rate of convergence is of paramount importance to prove conditions of type (31).

Concerning the problem of the Boltzmann equation with an energy source, Bobylev and Cercignani [BC02a] found steady solutions to the inelastic Maxwell model with a heat bath, that behave like $\exp(-r|v|)$. The problem of convergence towards the steady solution has been subsequently dealt with in [BCT05a]. By means of the contraction property of a suitable metric in the set of probability measures, existence, uniqueness, boundedness of moments and regularity of the steady state have been derived. Furthermore, explicit decay rates of general solutions towards the stationary state were obtained.

Using ideas from [BCT05a] in [BCT05b] various problems related to the convergence towards the self-similar profile were solved, reckoning precise rates of convergence in terms of suitable metrics defined in terms of the Fourier transform. Contraction properties of these metrics allow to show existence and uniqueness of the similarity solution, as well as various properties of the solution itself. In particular, it has been possible to discuss in detail the conjecture on the self-similar solution formulated by Ernst and Brito in [EB02a, EB02b].

A crucial role in the analysis of [BCT05b] is played by the weak norm convergence, which is obtained by further pushing the development of a method first used in [GTW95] to control the exponential convergence for Maxwellian molecules in certain weak norms.

Easy computations show that $(\varphi(v), Q(f, f)) = 0$ whenever $\varphi(v) = 1, v$, while $(\varphi(v), Q(f, f)) < 0$ if $\varphi(v) = v^2$. This corresponds to conservation of mass and momentum, and, respectively, to loss of energy for the solution to (32). For this reason, if we fix the initial data to be a centered probability density function, the solution will remain centered at any subsequent time $t > 0$.

Let $\mathcal{P}_s(I\!\!R^3)$ be the set of probability measures with bounded s-moment. The Fourier-based metrics d_s, for any $s > 0$, are defined as

$$d_s(\hat{f}, \hat{g}) = \sup_{k \in I\!\!R^3} \frac{|\hat{f}(k) - \hat{g}(k)|}{|k|^s}$$

for any pair of probability measures in $\mathcal{P}_s(I\!R^3)$. As usual, $\hat{f}$ is the Fourier transform of the density $f(v)$,

$$\hat{f}(k,t) = \int_{I\!R^3} f(v,t)\mathrm{e}^{-iv\cdot k}\,dv.$$

By simple Taylor expansion, one shows that the distance is well-defined and finite for any pair of probability measures with equal moments up to order $[s]$, where $[s]$ denotes the integer part of s. Moreover, in case $s \geq 1$ be an integer, it suffices equality of moments up to order $s - 1$ for being d_s finite. In fact, d_s with $s \geq 2$ topology is equivalent to the weak-star topology for measures plus convergence of moments up to order $[s]$ [TV99b], and can be related to the Wasserstein distance between probability measures. This distance can be considered as a Lyapunov functional in this case.

To fix ideas, let us set $\epsilon = 1$ in the Boltzmann equation (2), with the collision operator satisfying (32). The spatially homogeneous dissipative Boltzmann equation then reads

$$\frac{\partial f}{\partial t} = B\sqrt{\theta(t)}\tilde{Q}(f,f)\,. \tag{36}$$

Self-similar solutions of (32) are obtained through a suitable scaling of both time and velocity in such a way that energy of the solution is conserved. If

$$f(v,t) = \theta^{-\frac{3}{2}}(\tau)\,g(v\,\theta^{-\frac{1}{2}}(\tau),\tau),$$

where

$$\tau = \frac{B}{E}\int_0^t \theta^{\frac{1}{2}}(w)\,dw, \tag{37}$$

and $E = 8/(1 - e^2)$, g satisfies the equation

$$\frac{\partial g}{\partial \tau} = -\nabla_v \cdot \left(v\,g(v)\right) + EQ(g,g). \tag{38}$$

This equation now may admit a nontrivial steady state, because the energy dissipative effects of the Boltzmann collision operator are balanced by the energy input coming from the term $-\nabla_v \cdot \left(v\,g(v)\right)$. Self-similar solutions of the original Boltzmann equation correspond to stationary solutions g_∞ of (38).

Let us consider the pressure tensor for the solutions $f(v,\tau)$ of (36), namely, for $i \neq j$ the quantity

$$p_{ij}(\tau) = \int_{I\!R^3} v_i v_j f(v,\tau)\,dv.$$

If $\hat{\Phi}(k,\tau)$ is defined as

$$\hat{\Phi}(k,\tau) = \begin{cases} -\dfrac{1}{2}\displaystyle\sum_{i\neq j} p_{ij}(\tau)k_i k_j & if\ \ |k| \leq 1 \\[2mm] 0 & if\ \ |k| > 1 \end{cases} \tag{39}$$

the following theorem holds:

Theorem 1 ([BCT05b]). *Let $g(v,t)$ denote a solution to (38) correspond-ing to the initial value $g(0)$ with unit mass and zero mean velocity, then the following exponential decay towards the corresponding steady state g_∞ holds:*

$$d_2(\hat{g}(\tau),\hat{g}_\infty) \leq \frac{C_{2,2+\alpha}}{\theta_0} \left[2\, d_{2+\alpha}(\hat{g}(0) - \hat{\Phi}(0),\hat{g}_\infty) + C_1\right]^{2/(2+\alpha)}$$
$$\times \exp\left\{-\frac{2}{2+\alpha}\, C(\alpha,e)\,\tau\right\} + \frac{C_2}{\theta_0}\exp\left\{-\frac{1+e}{1-e}\tau\right\}.$$

(40)

The previous theorem gives explicit rates of exponential convergence to the steady state of the d_s-distance. The rate of convergence towards the self-similar solution (homogeneous cooling state) is easily derived from the previous theorem by coming back to the original time variable t. The evolution equation for the temperature yields

$$\theta(t) = \left\{\theta_0^{-\frac{1}{2}} + \frac{1-e^2}{8}Bt\right\}^{-2},$$

hence time scaling (37) is nothing but

$$\tau = \log\left[1 + \frac{B}{E\theta_0^{-\frac{1}{2}}}\,t\right].$$

Therefore, to any exponential decay in the variable τ there corresponds an algebraic decay in t. From Theorem 1 we get the following estimate for the convergence of each solution $f(v,t)$ towards the homogeneous cooling state $f_\infty(t)$:

$$d_2(\hat{f}(t),\hat{f}_\infty(t)) \leq C_{2,2+\alpha}\left[2\, d_{2+\alpha}(\hat{f}(0) - \hat{\Phi}(0),\hat{f}_\infty(0)) + C_1\right]^{2/(2+\alpha)}$$
$$\times \left[1 + \frac{B}{E\theta_0^{-\frac{1}{2}}}\,t\right]^{-(2(1-A(\alpha,e))E)/(2+\alpha)} + C_2\left[1 + \frac{B}{E\theta_0^{-\frac{1}{2}}}\,t\right]^{-(3-e)/(1-e)}.$$

The previous estimate guarantees that the homogeneous cooling state for the dissipative Boltzmann equation for Maxwell molecules attracts all solutions with finite moments of order $2 + \delta$, so that it is highly reasonable to use this cooling state to close fluid dynamics equations.

Analogous results for the dissipative Boltzmann equation with rigid-spheres kernel are not available. Recent mathematical studies [MM05a, MM05b] however, proved existence (without uniqueness) and various prop-erties of the cooling state in this case.

5 Conclusions

In this chapter, we briefly discussed the passage to hydrodynamics for rapid granular flows, described at a kinetic level in terms of the dissipative Boltzmann equation. This problem requires new hints on the mathematical tools necessary to justify rigorously the target macroscopic equations. Two possibilities are discussed. The first is linked to the presence of low inelasticity in the system, which allows to use locally Maxwellian functions to close macroscopic equations. We are confident in this case the passage can be justified by using methods close to thats applicable to the elastic Boltzmann equation. The second requires small spatial variations, and makes use of the homogeneous cooling state to close macroscopic equations. The first step in the justification of this procedure requires a proof of stability of the cooling state with respect to a large class of physical data. Recent mathematical results for the simplified model of the Boltzmann equation for Maxwell molecules then show that this stability property holds with respect to a distance equivalent to the weak* convergence of measures [BCT03]. The general situation is still unclear, and a result analogous to that for Maxwell molecules looks completely hopeless for the moment.

References

[BC02a]   Bobylev, A.V., Cercignani, C.: Moment equations for a Granular Material in a Thermal Bath' J. Statist. Phys. **106**, 547 (2002)

[BC02b]   Bobylev, A.V., Cercignani, C.: Self-similar solutions of the Boltzmann equation and their applications. J. Statist. Phys. **106**, 1039 (2002)

[BC03]   Bobylev, A.V., Cercignani, C.: Self-similar asymptotics for the Boltzmann equation with inelastic and elastic interactions. J. Statist. Phys. **110**, 333 (2003)

[BCG00]   Bobylev, A.V., Carrillo, J.A. Gamba, I.: On some properties of kinetic and hydrodynamics equations for inelastic interactions. J. Statist. Phys. **98**, 743 (2000)

[BCP97]   Benedetto, D., Caglioti, E., Pulvirenti, M.: A kinetic equation for granular media. Mat. Mod. Numer. Anal., **31**, 615 (1997)

[BCT03]   Bobylev, A.V., Cercignani, C., Toscani, G.: Proof of an asymptotic property of self-similar solutions of the Boltzmann equation for granular materials. J. Statist. Phys. **111**, 403 (2003)

[BCT05a]   Bisi, M., Carrillo, J.A., Toscani, G.: Decay rates towards self-similarity for the Ernst-Brito conjecture on large time asymptotics of the inelastic Maxwell model. Preprint (2005), 301–331

[BCT05b]   Bisi, M., Carrillo, J.A., Toscani, G.: Decay rates towards self-similarity for the Ernst-Brito conjecture on large time asymptotics of the inelastic Maxwell model. (Preprint) (2005)

[BDKS98]   Brey, J.J., Dufty, J.W., Kim, C.S., Santos, A.: Hydrodynamics for granular flows at low density. Phys. Rew. E **58**, 4638 (1998)

[BGL91]   Bardos, C., Golse, F., Levermore, C.D.: Fluid dynamic limits of kinetic equations. I. Formal derivations. J. Statist. Phys. **63**, 323 (1991)

[BGL93]   Bardos, C., Golse, F., Levermore, C.D.: Fluid dynamic limits of kinetic equations. II. Convergence proofs for the Boltzmann equation. Comm. Pure Appl. Math. **46**, 667 (1993)

[BK00]   Ben-Naim, E., Krapivski, P.: Multiscaling in inelastic collisions. Phys. Rev. E **61**, R5 (2000)

[BK02]   Ben-Naim, E., Krapivski, P.: Nontrivial velocity distributions in inelastic gases. J. Phys. A **35**, L147 (2002)

[BMP02]   Baldassarri, A., Marconi, U.M.B., Puglisi, A.: Influence of correlations in velocity statistics of scalar granular gases. Europhys. Lett. **58**, 14 (2002)

[Bob88]   Bobylev, A.V.: The theory of the nonlinear spatially uniform Boltzmann equation for Maxwell molecules. Sov. Sci. Rev. C. Math. Phys. **7**, 111 (1988)

[BP00a]   Brilliantov, N.V., Pöschel, T.: Granular gases with impact-velocity dependent restitution coefficient. in *Granular Gases*, T. Pöschel, S. Luding Eds. Lecture Notes in Physics, Vol. **564**, Springer Verlag, Berlin, (2000)

[BP00b]   Brilliantov, N.V., Pöschel, T.: Self–diffusion in granular gases. Phys. Rev. E **61**, 1716 (2000)

[BP00c]   Brilliantov, N.V., Pöschel, T.: Velocity distribution in granular gases of viscoelastic particles. Phys. Rev. E **61**, 5573 (2000)

[BP00d]   Brilliantov, N.V., Pöschel, T.: Granular Gases - The early stage. (in: Miguel Rubi (ed.) *Coherent Structures in Classical Systems*, Springer, New York (2000)

[BP02]   Brilliantov, N.V., Pöschel, T.: Hydrodynamics of Granular Gases of viscoelastic particles. Phil. Trans. R. Soc. Lond. A **360**, 415 (2002)

[BP03a]   Brilliantov, N.V., Pöschel, T.: Kinetic integrals in the kinetic theory of dissipative gases. (in *Granular Gas Dynamics*, T. Pöschel, N.V. Brilliantov Eds. Lecture Notes in Physics, Vol. **624**, Springer Verlag, Berlin, (2003)

[BP03b]   Brilliantov, N.V., Pöschel, T.: Hydrodynamics and transport coefficients for dilute granular gases. Phys. Rev. E **67**, 61304 (2003)

[BST04]   Bisi, M., Spiga, G., Toscani, G.: Grad's equations and hydrodynamics for weakly inelastic granular flows. Phys. Fluids **16**, 4235 (2004)

[Cer95]   Cercignani, C.: Recent developments in the mechanism of granular materials. *Fisica Matematica e ingegneria delle strutture*, Pitagora Editrice, Bologna, (1995)

[CIP94]   Cercignani, C., Illner, R., Pulvirenti, M.: *The mathematical theory of dilute gases* Springer Series in Applied Mathematical Sciences, Vol. **106**, Springer–Verlag, (1994)

[CJMTU01]   Carrillo, J.A., Jüngel, A., Markowich, P., Toscani, G., Unterreiter, A.: Entropy dissipation methods for degenerate parabolic problems and generalized Sobolev inequalities. Monatsh. Math. **133**, 1 (2001)

[CK70]   Chapman, S., Cowling, T.G.: *The mathematical theory of non–uniform gases* Cambridge University Press, Cambridge, (1970)

[CT00]   Carrillo, J.A., Toscani, G.: Asymptotic L^1-decay of solutions of the porous medium equation to self-similarity. Indiana Univ. Math. J. **49**, 113 (2000)

[CV02]   Caglioti, E., Villani, C.: Homogeneous cooling states are not always good approximations to granular flows. Arch. Ration. Mech. Anal. **163**, 329 (2002)

[DLK95]   Du, Y., Li, H., Kadanoff, L.P.: Breakdown of hydrodynamics in a one–dimensional system of inelastic particles. Phys. Rev. Lett. **74**, 1268 (1995)

[Duf01]   Dufty, J.W.: Kinetic theory and Hydrodynamics for rapid granular flows–A perspective. cond–mat/0108444v1 (2001)

[EB02a]   Ernst, M.H., Brito, R.: High energy tails for inelastic Maxwell models. Europhys. Lett. **58**, 182 (2002)

[EB02b]   Ernst, M.H., Brito, R.: Scaling solutions of inelastic Boltzmann equation with over-populated high energy tails. J. Statist. Phys. **109**, 407 (2002)

[EP97]   Esipov, S., Pöschel, T.: The granular phase diagram. J. Stat. Phys. **86**, 1385 (1997)

[GK02]   Gorban, A.N., Karlin, I.V.: Hydrodynamics from Grad's equations: What can we learn from exact solutions. Ann. Phys. **11**, 783 (2002)

[GPT97]   Gabetta, E., Pareschi, L., Toscani G.: Relaxation schemes for nonlinear kinetic equations. SIAM J. Numer. Anal. **34**, 2168 (1997)

[Gra49]   Grad, H.: On the kinetic theory of rarefied gases. Comm. Pure Appl. Math. **2**, 331 (1949)

[GS95]   Goldshtein, A., Shapiro, M.: Mechanics of collisional motion of granular materials. Part 1. General hydrodynamic equations. J. Fluid Mech. **282**, 75 (1995)

[GTW95]   Gabetta, E., Toscani, G., Wennberg, W.: Metrics for Probability Distributions and the Trend to Equilibrium for Solutions of the Boltzmann Equation. J. Statist. Phys. **81**, 901 (1995)

[Haf83]   Haff, P. K.: Grain flow as a fluid-mechanical phenomenon. J. Fluid Mech. **134**, 401 (1983)

[JR85]   Jenkins, J.T., Richman, M.W.: Grad's 13-moment system for a dense gas of inelastic spheres. Arch. Rational Mech. Anal. **87**, 355 (1985)

[Kog69]   Kogan, M.N.: *Rarefied Gas Dynamics* Plenum Press, New York, (1969)

[KW76]   Krook, T., Wu, T.T.: Formation of Maxwellian tails. Phys. Rew. Lett. **36**, 1107 (1976)

[LS86]   Lun, C.K.K., Savage, S.B.: The effects of an impact velocity dependent coefficient of restitution on stresses developed by sheared granular materials. Acta Mech. **63**, 15 (1986)

[MM05a]   Mischler, S., Mouhot, C.: Cooling process for the inelastic Boltzmann equation for hard spheres. Part I: The Cauchy problem. To appear on J. Statist. Phys. (2005)

[MM05b]   Mischler, S., Mouhot, C.: Cooling process for the inelastic Boltzmann equation for hard spheres. Part II: Self–similar solutions and tail behavior. To appear on J. Statist. Phys. (2005)

[NY93]   McNamara, S., Young, W.R.: Kinetics of a one–dimensional granular medium in the quasi–elastic limit. Phys. Fluids A **5**, 34 (1993)

[PR01]   Pareschi, L,, Russo, G.: Time relaxed Monte Carlo methods for the Boltzmann equation. SIAM J. Sci. Comput. **23**, 1253 (2001)

[RC02]   Risso, D., Cordero, P.: Dynamics of rarefied granular gases. Phys. Rew. E **65**, 21304 (2002)

[RPBS99]   Ramírez, R., Pöschel, T., Brilliantov, N.V., Schwager, T.T.: Coefficient of restitution of colliding viscoelastic spheres. Phys. Rev. E **60**, 4465 (1999)

[SP98]   Schwager, T.T., Pöschel, T.: Coefficient of normal restitution of viscous particles and cooling rate of granular gases. Phys. Rev. E **57**, 650 (1998)

[Tos04]   Toscani, G.: Kinetic and hydrodynamic models of nearly elastic granular flows. Monatsc. Math. **142**(1–2), 179 (2004)

[TV99a]   Toscani, G., Villani, C.: Sharp entropy production bounds and explicit rate of trend to equilibrium for the spatially homogeneous Boltzmann equation. Commun. Math. Phys. **203**, 667 (1999)

[TV99b]   Toscani, G., Villani, C.: Probability Metrics and Uniqueness of the Solution to the Boltzmann Equation for a Maxwell Gas. J. Statist. Phys. **94**, 619 (1999)

[Vaz83]   Vázquez, J.L.: Asymptotic behavior and propagation properties of the one-dimensional flow of a gas in a porous medium. Trans. Amer. Math. Soc. **277**, 507 (1983)

[Vaz03]   Vázquez, J.L.: Asymptotic behaviour for the porous medium equation in the whole space. J. Evol. Equ. **3**, 67 (2003)

Bodies with Kinetic Substructure

Gianfranco Capriz

Summary. In some earlier papers an elementary approach was followed to suggest a set of balance laws governing, within a continuum theory, the evolution of bodies made up of countless molecules afflicted by chaotic agitation. The set is larger than usual to insure strict observer independence of consequent thermal entities. Here preliminary steps are taken to pursue the same goal but with an inception closely akin to that prefacing the kinetic theory of gases; the quest here, however, exacts divergence from the route followed in the latter theory. Thus some, possibly controversial, notions emerge and are proffered here for criticism.

1 Kinetics

Consider a body in its deportment at an instant τ, when it occupies a region $\mathcal{B}$; a region which, in imagination, is envisaged as split into tiny spatial segments. Each segment $\mathfrak{e}$ contains many molecules and, although it is said to be located at a place x within $\mathcal{B}$, it must be imagined to have a microexpanse within which subplaces can be distinguished at a lower scale. Accordingly, and contrary to the bias mooted by the standard kinetic theory, of each molecule one presumes here to gauge not only the velocity w (which in principle can be any member of the vector space $\mathcal{V}$) but also the subplace z within $\mathfrak{e}$ (z being distinct, at our penetrating magnification, from x).

Consequently one seeks the distribution θ, valid for $\mathfrak{e}(x)$ at time τ, such that $\theta(\tau, x; z, w)\, dzdw$ gives the number of molecules passing in the vicinity of z and with velocity near w. θ is presumed to be such that all integrations involving it and mentioned below are convergent. In particular

$$\omega = \int_{\mathfrak{e}} \int_{\mathcal{V}} \theta(\tau, x; z, w), \qquad [\theta] = L^{-6}T^3, \tag{1}$$

gives the (finite, though large) number of molecules in $\mathfrak{e}(x)$ at time τ. Take note that, at any instant τ, there may be many molecules passing through the immediate neighbourhood of z, possibly with widely different velocities.

If all molecules have the same mass μ (as always presumed below for simplicity) then

$$\mu\omega = (meas\ \mathfrak{e})\,\rho, \tag{2}$$

where ρ is the gross mass density at x.

Some formulae below are shortened by use of the distribution $\tilde{\theta}$

$$\tilde{\theta}\,(\tau, x; z) = \int_{\mathcal{V}} \theta\,(\tau, x; z, w)\,, \qquad \left[\tilde{\theta}\right] = L^{-3}, \tag{3}$$

which counts the number of molecules near z whatever their velocity.

Vice versa, within the kinetic theory, as already mentioned, only the alternative reduced distribution $\hat{\theta}$ matters,

$$\hat{\theta}\,(\tau, x; w) = \int_{\mathfrak{e}} \theta\,(\tau, x; z, w)\,, \qquad \left[\hat{\theta}\right] = L^{-3}T^{3}, \tag{4}$$

which counts the number of molecules in the whole $\mathfrak{e}$ and velocity near w.

Using θ, or $\tilde{\theta}$, one determines he centre of gravity of all molecules in $\mathfrak{e}$

$$x = \omega^{-1} \int_{\mathfrak{e}} \int_{\mathcal{V}} \theta z = \omega^{-1} \int_{\mathfrak{e}} \tilde{\theta} z; \tag{5}$$

it is after such x that the segment is labelled. Then z can be split into x and y, and, with a slight abuse of notation, one has

$$\int_{\mathfrak{e}} \int_{\mathcal{V}} \theta y = \int_{\mathfrak{e}} \tilde{\theta} y = 0. \tag{6}$$

As it is done, with success, in the standard theory of fluids, the velocity v assigned at x is, by fiat, the average velocity

$$v = \omega^{-1} \int_{\mathfrak{e}} \int_{\mathcal{V}} \theta w. \tag{7}$$

Similarly one attributes to the subelement at z the average velocity $\tilde{w}$ of all molecules passing there

$$\tilde{w} = \tilde{\theta}^{-1} \int_{\mathcal{V}} \theta w \tag{8}$$

so that, in particular,

$$v = \omega^{-1} \int_{\mathfrak{e}} \tilde{\theta}\tilde{w}, \qquad \omega = \int_{\mathfrak{e}} \tilde{\theta}. \tag{9}$$

Thus correct evaluation of total momentum for $\mathfrak{e}$ is assured summoning reduced quantities $\tilde{\theta}$ and $\tilde{w}$ only. Actually v can be secured also by turning to $\hat{\theta}$ only

$$v = \omega^{-1} \int_{\mathcal{V}} \hat{\theta} w, \qquad \omega = \int_{\mathcal{V}} \hat{\theta}, \tag{10}$$

as in the kinetic theory of gases, within which, though, no meaning can be attached to $\tilde{w}$.

2 A Shadow Speck of Matter

Availability of the fields v and $\tilde{w}$ grants us licence to invent a shadow speck of matter which, in imagination, simply translates with velocity v and within which, besides, the shadow subspeck at z flies with relative velocity $\tilde{w}$. We can then deal with the speck as it were a subbody (in the sense of standard theory of continua) rather than a collection of riotous molecules, a subbody occupying instantaneously the segment $\mathfrak{e}$ consisting of subplaces each identified by the variable z, where the material density is $\mu\tilde{\theta}$.

One may now proceed to evaluate *Euler's inertia tensor* Y around x,

$$Y = \omega^{-1} \int_{\mathfrak{e}} \int_{\mathcal{V}} \theta y \otimes y = \omega^{-1} \int_{\mathfrak{e}} \tilde{\theta} y \otimes y \tag{11}$$

and the tensor *moment of momentum* K

$$K = \omega^{-1} \int_{\mathfrak{e}} \int_{\mathcal{V}} \theta y \otimes (w - v) = \omega^{-1} \int_{\mathfrak{e}} \int_{\mathcal{V}} \theta y \otimes w = \omega^{-1} \int_{\mathfrak{e}} \tilde{\theta} y \otimes \tilde{w}. \tag{12}$$

Neither tensor could be defined with access to the distribution $\tilde{\theta}$ only. Vice versa the kinetic energy tensor per unit mass W

$$W = \frac{1}{2}\omega^{-1} \int_{\mathfrak{e}} \int_{\mathcal{V}} \theta w \otimes w = \frac{1}{2}\omega^{-1} \int_{\mathcal{V}} \hat{\theta} w \otimes w \tag{13}$$

cannot be achieved with the distribution $\tilde{\theta}$ only. Thus, the 'reduced' tensor $\tilde{W}$ acquires a decisive reserve rôle

$$\tilde{W} = \frac{1}{2}\omega^{-1} \int_{\mathfrak{e}} \int_{\mathcal{V}} \theta \tilde{w} \otimes \tilde{w} = \frac{1}{2}\omega^{-1} \int_{\mathfrak{e}} \tilde{\theta} \tilde{w} \otimes \tilde{w}. \tag{14}$$

The difference

$$W - \tilde{W} = \frac{1}{2}\omega^{-1} \int_{\mathfrak{e}} \int_{\mathcal{V}} \theta (w - \tilde{w}) \otimes (w - \tilde{w}) \tag{15}$$

will be relegated within some 'internal energy' tensor.

One can take now a further step and invent for the shadow speck a congruent affine kinetic field with a rate of deformation B, say; congruent in the sense that, for it, the tensor moment of momentum, now amounting to YB^T, is still equal to K: B need only be chosen to coincide with $K^T Y^{-1}$. There is a similarity here with the process that led to the selection of v: in that case it was the global momentum of the molecules pertaining to the segment which turned out to be equal to that which would have been experienced, within the segment, were all molecules to fly with the same velocity v. In the devised affine impetus, the tensor moment of momentum remains that occurring in the real molecular transit through $\mathfrak{e}$.

Having also assembled the field $B(x, \tau)$, one can imagine it generated by a fictitious affine deformation G from an arbitrary constant reference. In principle one need only integrate the partial differential equation

$$\frac{\partial G}{\partial \tau} + (grad G)\, v = BG, \tag{16}$$

an integration which determines G a constant right factor apart. Basically the process in not different from that which leads to trajectories in ordinary fluid dynamics through an integration over v (a vector which we know to be an average over a population, not the property of a specific mass-point).

Abiding by the notation $\tilde{N} = GG^T$ and $R' = \tilde{N}^{-\frac{1}{2}}G$ used in an earlier paper [1] G can be split into the product

$$G = \tilde{N}^{\frac{1}{2}}R', \tag{17}$$

with the orthogonal tensor R' providing an intrinsic local reference $\mathcal{R}$. The inverse G^{-1} could be intended to express the retrogression of $\mathfrak{e}$ into a paragon segment $\mathfrak{e}_*$ and of the subplace y into a paragon subplace s: $y = Gs$ (in such a way, we recall, that K does not change if, in its definition, $w - v$ is substituted by $\dot{G}s$).

The average molecular velocity with respect to $\mathcal{R}$, $\tilde{w} - v - \dot{G}s$, can be pulled back with the help of G to provide us with the 'peculiar' velocity c

$$c = G^{-1}\left(\tilde{w} - v - \dot{G}s\right). \tag{18}$$

Some additional remarks:

(i) c is observer-independent; any rotation of the observer does not influence the reading of c.

(ii) The choices of s and c are such that, not only $\int_{\mathfrak{e}} \tilde{\theta}s = 0$, $\int_{\mathfrak{e}} \tilde{\theta}c = 0$, but also

$$\int_{\mathfrak{e}} \tilde{\theta}s \otimes c = 0. \tag{19}$$

(iii) Those choices make the integral

$$\int_{\mathfrak{e}} \tilde{\theta}\left(\tilde{w} - v - By\right)^2 \tag{20}$$

a minimum; thus, also in this sense, the option suggested for B is best fitting.

A crucial precondition for progress is to make it clear, even if repetitive, that the spatial segment $\mathfrak{e}$ is meant to be interpreted as the instantaneous placement of a fictitious material speck which translates with the velocity v and deforms affinely with a rate directed by B; the placement $\mathfrak{e}$ derives from another changeless fictional placement $\mathbf{e}^*$, with the central assumption that the transplacement from $\mathbf{e}^*$ to $\mathbf{e}$ preserves mass:

$$\overline{\mu\theta \det G} = 0. \tag{21}$$

A decisive corollary ensues: if $\mathcal{G}$ is a sufficiently regular function of z and τ, then

$$\left(\int_{\mathbf{e}} \tilde{\theta}\mathcal{G}\right)^{\cdot} = \int_{\mathbf{e}} \tilde{\theta}\mathcal{G}^{\cdot}. \tag{22}$$

Hence, in particular, $\dot{\omega} = 0$.

The assumption (21) is compatible with macroscopic mass balance because molecules of one speck may protrude into and from neighbouring specks. Inside $\mathbf{e}$ the agitation of the molecules is described only within the limits allowed by the assignment of the field $\tilde{w}$. A global estimate of the intensity of agitation at x (i.e., within $\mathbf{e}$) is offered by the tensor H:

$$H = \omega^{-1} \int_{\mathbf{e}} \tilde{\theta} \left(\tilde{w} - v - \dot{G}s\right) \otimes \left(\tilde{w} - v - \dot{G}s\right) = \omega^{-1} G \left(\int_{\mathbf{e}} \tilde{\theta} c \otimes c\right) G^T; \tag{23}$$

notice that

$$\tilde{W} = BYB^T + H. \tag{24}$$

Exploiting the shadow kinetics, one finds that

$$\dot{Y} = \left(\omega^{-1} \int_{\mathbf{e}} \tilde{\theta} y \otimes y\right)^{\cdot} = \omega^{-1} \int_{\mathbf{e}} \overline{\tilde{\theta}(y \otimes y)}^{\cdot} =$$

$$= \omega^{-1} \int_{\mathbf{e}} \tilde{\theta} \left(w \otimes y + y \otimes w\right) = K^T + K = BY + YB^T. \tag{25}$$

On the other hand from the equation for G above one gets also

$$\left(GG^T\right)^{\cdot} = BGG^T + GG^T B^T. \tag{26}$$

Thus the 'strain' GG^T satisfies the same condition required of Y; choosing the arbitrary factor so as to adjust also dimensions one is led to the identification

$$Y = (meas \; \mathbf{e})^{\frac{2}{3}} GG^T. \tag{27}$$

Y can be interpreted as an intrinsic metric at x and, ultimately at all points occupied by the body.

3 Straining and Allied Notions

It may be argued that our entire analysis balances precariously on the razor edge of ingrained ambiguities tied with the simultaneous concerns with two scales; misconceptions must be prevented already with regards to the notion of straining.

Above the tensor G was sought from knowledge of B; likewise the formal construction of the placement gradient F can be effected. However, whereas the former rendition is conceived strictly within $\mathfrak{e}$, the latter demands knowledge of v over all elements in the immediate gross neighbourhood of x, so that $L = grad_x v$ be available. Then F can be sought as a solution of $\dot{F} = FL$ and there is no geometric reason for F to be conditioned by G, nor, of course, L by B. Furthermore, within $\mathfrak{e}$, one can evaluate $grad_y \tilde{w}$:

$$grad_y \tilde{w} = B + G\,(grad_y c), \tag{28}$$

yet another distinct tensor, which averaged over $\mathfrak{e}$

$$\omega^{-1} \int_{\mathfrak{e}} \tilde{\theta}\, grad_y \tilde{w} = B + \omega^{-1} G \left(\int_{\mathfrak{e}} \tilde{\theta}\, grad_y c \right) \tag{29}$$

leads to a new field over $\mathcal{B}$, say $J(x, \tau)$, which assesses a sort of micro-stretching and spin evoked from the molecular maelstrom and to be, possibly, attributed to $\mathfrak{e}$.

Thus B, chosen to estimate most fittingly the relative kinetic energy is not quite as successful in matching average micro straining. Of course, the additional term might still vanish or, at least, amount to little and thus be negligible; only the scrutiny of many special instances will offer evidence one way or another. Rewriting the correction to B in the form

$$\omega^{-1} G \left(\int_{\mathfrak{e}} \tilde{\theta}\, grad_y c \right) = \omega^{-1} G \left(\int_{\partial \mathfrak{e}} \tilde{\theta} c \otimes n - \int_{\mathfrak{e}} c \otimes grad_y \tilde{\theta} \right) \tag{30}$$

(n the normal to $\partial \mathfrak{e}$) evidences a contribution due to a flux through $\partial \mathfrak{e}$ and one due to a rearrangement within $\mathfrak{e}$.

Below attention is focused on the requited rôle of F versus G or of L versus B. In a sense, G may be envisioned to account for:

(i)   The influence within the element of the macrostretch F, plus
(ii) The rearrangement of molecules within the macrostretched element insofar as a crowding near the centre implies a smaller moment of inertia than a crowding at the periphery, and
(iii) The protrusion of molecules beyond the element bounds after they are expanded by the macrostretch and insofar as they can be accounted for affinely.

Above the concepts of stretch are, of course, virtual as quantities derived from an irregularly evolving reality. They might, nevertheless, take up direct capacity within some ensuing developments; then the formal splitting of G into the product of GF^{-1} by F (or, rather, of F by $F^{-1}G$) separates nominally the outcome of action (i) from the other two; to the combined effect of the latter the contribution of (ii) could be measured by GG^T though such choice includes consequences of protrusion proper (though excluding, however, the effects mentioned in the previous paragraph).

In an essay on perfect pseudofluids [2], the following strain characteristics were invoked

$$C = F^T F, \quad N = G^T G, \quad X = G^{-1} F, \tag{31}$$

leading to the rates

$$\dot{C} = 2F^T \left(sym L\right) F, \quad \dot{N} = 2G^T \left(sym B\right) G, \quad \dot{X} = G^{-1} \left(L - B\right) F. \tag{32}$$

Notice that $\dot{X}$ is not independent of $\dot{C}$, $\dot{N}$. Thus, strictly, X is not the appropriate characteristic to pool with C and N; rather that rôle could be properly taken by

$$Q = R'^T R, \tag{33}$$

where R and R' are the orthogonal tensors associated with F and G respectively, with

$$F = R C^{\frac{1}{2}}, \quad G = R' N^{\frac{1}{2}}. \tag{34}$$

In fact,

$$\dot{Q} = R'^T \left(\dot{R} R^T - \dot{R'} R'^T \right) R = R'^T \left(skw L - skw B\right) R \tag{35}$$

is evidently independent of $\dot{C}$, $\dot{N}$.

Protrusion does not necessarily mean loss or gain of molecules: in an element number density may easily be balanced by intrusion from neighbouring elements. Thus a discrepancy between F and G by itself is insufficient to imply mass variation, it might simply give a hint as to the extent of interpenetration. A scalar measure of the latter could be the different change of volume attributed by F and G: $\alpha = \det \left(F G^{-1}\right)$, leading to the rate

$$\dot{\alpha} = \alpha\, tr \left(L - B\right). \tag{36}$$

Rather, it is only in the presence of a relatively steep gradient of α or, more generally, of X that protrusion implies deviant features. Thus that gradient enters necessarily among descriptive variables, perhaps through associated quantities, such as wryness, torsion, Burgers' vector, but also 'extra matter'.

Strain measures like C, N and Q appear inappropriate when addressing phenomena in fluids; in fact one may deem the bare pull-back linked to F, or G, as artificial; although a reference state could still be imagined: e.g., one where molecules are distributed homogenously within the element at some standard number density. Also, one must not disregard the opportunity offered apparently by those strain measures, to compare and contrast models of semisolids subject to 'configurational' changes, i.e. to mutations of background.

Strictly, when seeking theories for fluids, one should rather evidence measures bearing only on the current state such as the metrics

$$\tilde{C} = F F^T \quad \text{and} \quad \tilde{N} = \left(meas\ \mathfrak{e}\right)^{-\frac{2}{3}} Y = G G^T, \tag{37}$$

the rotation

$$\tilde{Q} = R'R^T,$$ (38)

a *wryness* w defined as the gradient of GF^{-1}, the consequent *torsion* h, dislocation density and Burgers' vector b (relative to any plane of normal n) given by

$$w = grad\left(GF^{-1}\right), \quad h = \frac{1}{2}\left(w - w^t\right), \quad b = \left(eh^T\right)n,$$ (39)

where the exponents t and T mean minor right transposition and major transposition respectively in the third-order tensors w and h, e is Ricci's permutation tensor. The common invariants of all those tensors have then a crucial rôle to play.

4 Balance Laws

The scenario promoted in the previous sections evidences within the region $\mathcal{B}$, once totals over each $\mathfrak{e}$ are affected, the substratum provided by the fields of gross density ρ and moment of inertia Y

$$\rho = \frac{\mu}{(meas\ \mathfrak{e})}\int_{\mathfrak{e}}\tilde{\theta}, \quad Y = \omega^{-1}\int_{\mathfrak{e}}\tilde{\theta}y \otimes y$$ (40)

and, later, the kinematic fields v, B and H

$$v = \omega^{-1}\int_{\mathfrak{e}}\tilde{\theta}\tilde{w}, \qquad B = \left(\int_{\mathfrak{e}}\tilde{\theta}\tilde{w}\otimes y\right)\left(\int_{\mathfrak{e}}\tilde{\theta}y\otimes y\right)^{-1},$$ (41)

$$H = \omega^{-1}\int_{\mathfrak{e}}\tilde{\theta}\left(\tilde{w} - v - By\right)\otimes\left(\tilde{w} - v - By\right).$$ (42)

Thus, the intention is not to press the depth of description of events in the body down to the details of the distribution θ (or, yet less deeply, $\tilde{\theta}$) but to stop at the stage set by those fields. Further, one expects that the evolution of the latter be ruled by balance laws also lingering at their level, hence involving, on the one hand, the time derivatives of v, B (or, better, K), H and, on the other hand, totals over $\mathfrak{e}$ of impact and/or bonding effects be those intimate (or close, i.e. among subspecks within $\mathfrak{e}$), internal (among distinct specks), external to the body. Such totals per unit mass are formally expressed by the integrals

$$\omega^{-1}\int_{\mathfrak{e}}\tilde{\theta}g^c, \quad \omega^{-1}\int_{\mathfrak{e}}\tilde{\theta}g^i, \quad \omega^{-1}\int_{\mathfrak{e}}\tilde{\theta}g^e$$ (43)

and could, in principle, be given substance once a collision/coherence operator (as occurs in Boltzmann equation) were known. Be that as it may, the presumption below is that $\tilde{w}$ equals the sum $g^c + g^i + g^e$ as per Newton law.

Disregarding, as said above, possible deeper inhomogeneities (which would be gauged by $\tilde{\theta}(\tau, z)$ and would be related to $div\ \tilde{w}$), *conservation of mass* is invoked by the standard law

$$\dot{\rho} + \rho div\ v = 0. \tag{44}$$

What could be called *law of conservation of moment of inertia* was already written (see (27)) and follows immediately from the definition (40)$_2$

$$\dot{Y} = 2symK. \tag{45}$$

Because totals of intimate interactions vanish, *conservation of momentum* embodied by (see (41)$_1$ and (43))

$$\rho\dot{v} = \rho\omega^{-1}\int_e \tilde{\theta}\left(g^i + g^e\right), \tag{46}$$

might take the usual form

$$\rho\dot{v} = \rho b + divT, \tag{47}$$

though here one should justify anew the presumption that external actions sum up into a functional absolutely continuous with gross volume, whereas internal actions obey Cauchy's assertions.

Conservation of moment of momentum follows from the definition of K (see (12) and, again, (43))

$$\dot{K} = \omega^{-1}\int_e \tilde{\theta}\tilde{w}\otimes\tilde{w} + \omega^{-1}\int_e \tilde{\theta}y\otimes\left(g^c + g^i + g^e\right), \tag{48}$$

from the link (24) and the property (25)

$$\omega^{-1}\int_e \tilde{\theta}\tilde{w}\otimes\tilde{w} = \omega^{-1}\int_e \tilde{\theta}\left(\dot{G}s + Gc\right)\otimes\left(\dot{G}s + Gc\right) = BK + H, \tag{49}$$

with the conclusion

$$\dot{K} - BK - H = \omega^{-1}\int_e \tilde{\theta}y\otimes\left(g^c + g^i + g^e\right). \tag{50}$$

Notation introduced in earlier papers could be called upon

$$M = \omega^{-1}\int_e \tilde{\theta}y\otimes g^e, \qquad A = -\frac{1}{\rho}\int_e \tilde{\theta}y\otimes g^c. \tag{51}$$

No impelling case, but analogy and convenience, is yet available to declare that the third addendum in the right-hand side of (50) be expressible as the divergence of a third-order tensor m, the factor ρ^{-1} apart, again as used in earlier papers. But, when that is the case the next balance law reads

$$\rho \left(\dot{K} - BK - H \right) = \rho M - A + div\ \mathsf{m}. \tag{52}$$

Finally one finds, again with reference to (43), (24), (25)

$$\dot{H} = 2sym \left[\omega^{-1} \int_{\mathfrak{e}} \tilde{\theta} \left(g^c + g^i + g^e - \dot{G}c \right) \otimes Gc \right] \tag{53}$$

or

$$\dot{H} + 2symBH = 2\omega^{-1} sym \int_{\mathfrak{e}} \tilde{\theta} \left(g^c + g^i + g^e \right) \otimes Gc. \tag{54}$$

Again, using notation of earlier papers for tensor virials

$$S = 2\omega^{-1} sym \int_{\mathfrak{e}} \tilde{\theta} g^e \otimes Gc, \tag{55}$$

$$Z = -\frac{1}{\rho} sym \int_{\mathfrak{e}} \tilde{\theta} g^c \otimes Gc, \tag{56}$$

and presuming again that also the virial of internal actions have contact character so that they be expressed as the divergence of a third-order tensor s, the last balance equation takes the disguise

$$\rho \left(\dot{H} + 2symBH \right) = \rho S - Z + div\mathsf{s}. \tag{57}$$

5 Balance of Kinetic Energy

Energy has the leading rôle in the continuum discussed here. Thus it seems appropriate to assemble a few results below, even if largely mentioned elsewhere.

Within our model the kinetic energy tensor per unit mass W can be split thus

$$W = \tilde{W} + U \tag{58}$$

with a thermal contribution

$$U = \frac{1}{2}\omega^{-1} \int_{\mathfrak{e}} \int_{\mathcal{V}} \theta \left(w - \tilde{w} \right) \otimes \left(w - \tilde{w} \right) \tag{59}$$

and a properly kinetic one

$$\tilde{W} = \frac{1}{2\omega} \int_{\mathfrak{e}} \tilde{\theta} \tilde{w} \otimes \tilde{w} = \frac{1}{2\omega} \int_{\mathfrak{e}} \tilde{\theta} \left(v + \dot{G}s + Gc \right) \otimes \left(v + \dot{G}s + Gc \right), \tag{60}$$

or, see remarks at the end of Sect. 2,

$$\tilde{W} = v \otimes v + BYB^T + H. \tag{61}$$

The tensorial kinetic energy theorem follows from the balance equations (47), (52), (57) multiplying tensorially the first by v, the second by B, summing the two with the third one (divided by 2) term by term, taking the symmetric part of both sides and integrating, by parts where appropriate, over the region occupied by the body

$$\int_B \rho \ddot{W} = \int_B \rho sym \left(v \otimes f + BM + \frac{1}{2}S \right) - $$
$$- \int_B \left(sym \left(\frac{1}{2}Z + LT^T \right) + BA + \mathsf{b}\mathsf{m}^t \right) + \tag{62}$$
$$+ \int_{\partial B} sym \left(v \otimes Tn + B\left(\mathsf{m}n \right) + \frac{1}{2}\mathsf{s}n \right),$$

where n is the unit normal vector to ∂B, b is the gradient of B, and the exponent t to m indicates minor right transposition: $\left(\mathsf{b}\mathsf{m}^t \right)_{ij} = B_{ia,b}\mathsf{m}_{ajb}$. The central term in the right-hand side must be interpreted as the tensor power of intimate and internal actions, with densities respectively

$$-sym \left(\frac{1}{2}Z + BA \right) \quad \text{and} \quad -sym \left(LT^T + \mathsf{b}\mathsf{m}^t \right). \tag{63}$$

Hence the density of scalar power is given by

$$-\left(\frac{1}{2}trZ + L \cdot T + B \cdot A^T + \mathsf{b} \cdot \left(\mathsf{m}^t \right)^T \right). \tag{64}$$

The equation of balance of moment of momentum (52) does not secure here observer independence of (64), as occurs in the classical case for the vectorial version. Two observers on frames in relative motion read different values of L and B: the change in both is the addition of the same skew tensor. Hence observer independence is assured if and only if

$$skw T = skw A. \tag{65}$$

If one were to demand observer independence of the tensor power then the stronger condition

$$T = -A^T \tag{66}$$

would be required, when the tensor power would reduce to

$$-sym \left(\frac{1}{2}Z + (L - B)T^T + \mathsf{b}\mathsf{m}^t \right). \tag{67}$$

It was already remarked in Sect. 3 that $symL$ and $symB$ can be expressed in terms of the strain rates $\dot{C}$ and $\dot{N}$ respectively and $skw\,(L-B)$ in terms of $\dot{Q}$

$$symL = \frac{1}{2}F^T\dot{C}F^{-1}, \qquad symB = \frac{1}{2}G^T\dot{N}G^{-1},$$

$$skw\,(L-B) = R'\dot{Q}R^T. \tag{68}$$

Longer algebra shows that

$$b_{ijk} + b_{jik} = G^{-1}_{Bi}\dot{n}_{ABK}G^{-1}_{Aj}F^{-1}_{Kk} - \dot{N}_{AB}G^{-1}_{Ba}G_{aC,k}\left(G^{-1}_{Ai}G^{-1}_{Cj} + G^{-1}_{Aj}G^{-1}_{Ci}\right) \tag{69}$$

where $n = (gradN)\,F$.

Thus the scalar power density can be written as an affine function of $\dot{C}, \dot{N}, \dot{Q}, \dot{n}$

$$-trZ - \left({}^tb - b\right)\cdot m - \left(F^{-1}\left(symT\right)F^{-T}\right)\cdot\dot{C} - \left(G^{-1}\left(symA\right)G^{-T}\right)\cdot\dot{N} +$$

$$+ \left(G^{-1}_{Ai}G^{-1}_{Cj} + G^{-1}_{Aj}G^{-1}_{Ci}\right)G^{-1}_{Ba}G_{aC,k}m_{ijk}\dot{N}_{AB} - \tag{70}$$

$$- 2\left[R'^T\left(skwT\right)R\right]\cdot\dot{Q} - G^{-1}_{Ai}G^{-1}_{Bj}m_{ijk}F^{-1}_{Ck}\dot{n}_{ABC}.$$

This result suggests the possible existence of continua for which a potential $\varphi\left(C, N, Q, m\right)$ exists and is such that

$$symT = 2\rho F\frac{\partial\varphi}{\partial C}F^T, \qquad skwT = skwA = \rho R'\frac{\partial\varphi}{\partial Q}R^T, \tag{71}$$

$$m_{ijk} = 2\rho G_{iA}G_{jB}F_{kC}\frac{\partial\varphi}{\partial n_{ABC}}; \tag{72}$$

thus m is symmetric in the first two indices and, as a consequence, the second term in the sum (70) vanishes. The factor multiplying $\dot{N}_{AB}$ is equal to

$$G^{-1}_{Ai}\left(symA\right)_{ij}G^{-1}_{Bj} + 2\rho G_{iR}G_{jS}F_{kT}\frac{\partial\varphi}{\partial n_{RST}}\left(G^{-1}_{Ai}G^{-1}_{Cj} + G^{-1}_{Aj}G^{-1}_{Ci}\right)G^{-1}_{Ba}G_{aC,k} \tag{73}$$

and hence

$$\left(symA\right)_{ij} = 2\rho G_{iA}\frac{\partial\varphi}{\partial N_{AB}}G_{jB} - 2\rho\left(G_{iR}G_{jS}\right)_k F_{kT}\frac{\partial\varphi}{\partial n_{RST}}. \tag{74}$$

Finally

$$trZ = 2\rho\dot{\varphi}. \tag{75}$$

When the constitutive laws above apply, the balance equations of momentum and tensor moment of momentum acquire the rôle of evolution equations for v and B (or x and G). The rule of progress for H needs additional physical insight.

6 The First Principle

A deeper kinetic energy theorem ensues if molecular events are graded more finely inside the distribution θ rather than $\tilde{\theta}$. Then, some intriguing corollaries ensue; their deduction is barely sketched below omitting adscititious qualifications to display the essence.

Choose $\theta(z, w) h \, dw dz$ to represent the resultant of the forces acting on the molecules belonging to the immediate neighbourhood of z, w, molecules numbering $\theta dw dz$ and h to be eventually split into the sum $h^c + h^i + h^e$, as g was earlier.

Then $\dot{w} = h$ and

$$\dot{W} = \left(\frac{1}{2} \omega^{-1} \int_e \int_\mathcal{V} \theta w \otimes w \right)^{\cdot} = \omega^{-1} \int_e \int_\mathcal{V} \theta sym \,(w \otimes h). \tag{76}$$

Recall notation introduced at the beginning of Sect. 5

$$W = \tilde{W} + U. \tag{77}$$

Hence

$$\dot{\tilde{W}} + \dot{U} = \omega^{-1} \int_e \int_\mathcal{V} \theta sym \,(w \otimes (h - g)) + \omega^{-1} \int_e \int_\mathcal{V} \theta sym \,(w \otimes g); \tag{78}$$

but, from the restricted kinetic energy theorem and the appropriate interpretation of terms

$$\int_\mathcal{B} \rho \dot{\tilde{W}} = \int_\mathcal{B} \mu \,(meas \; \mathfrak{e})^{-1} \int_e \int_\mathcal{V} \theta sym \,(w \otimes g) - $$
$$- \int_\mathcal{B} sym \left(\frac{1}{2} Z + LT^T + BA + \mathsf{bm}^t \right) \tag{79}$$

so that

$$\int_\mathcal{B} \rho \dot{U} = \int_\mathcal{B} \mu \,(meas \; \mathfrak{e})^{-1} \int_e \int_\mathcal{V} \theta sym \,(w \otimes (h - g)) + $$
$$+ \int_\mathcal{B} sym \left(\frac{1}{2} Z + LT^T + BA + \mathsf{bm}^t \right). \tag{80}$$

Finally, through the standard criterion of localization justified by the fact that the law above would equally apply when the integrals were extended to any subbody of $\mathcal{B}$,

$$\rho \dot{U} = \mu \,(meas \; \mathfrak{e})^{-1} \int_e \int_\mathcal{V} \theta sym \,(w \otimes (h - g)) + $$
$$+ sym \left(\frac{1}{2} Z + LT^T + BA + \mathsf{bm}^t \right). \tag{81}$$

Such is the local equation which expresses, under the circumstances, the first principle of thermodynamics.

Acknowledgement

This research was supported by the Italian Ministry MIUR through the project Mathematical Models for Materials Science. The sources of the ideas developed here are obvious (from the classical tracts to contributions dedicated to recent debates, e.g., by L.C. Woods); consequently they are not quoted in the references where only a couple of strictly connected papers are listed.

References

1. Capriz, G., Muellenger, G.: A Theory of Perfect Hyperfluids. In: Trends in Applications of Mathematics to Mechanics. Proceedings of STAMM, Seeheim, Germany, Aachen: Shaker, Berichte aus der Mathematik, 85–92 (2005)
2. Capriz, G.: Pseudofluids. In: Capriz, G., Mariano, P.M. (eds) Material Substructures in Complex Bodies: from Atomic Level to Continuum. Elsevier Science B.V., Amsterdam, 238–261 (2007)

From Extended Thermodynamics to Granular Materials

Tommaso Ruggeri

Summary. Taking into account the analogy with the kinetic approach of rarefied gases, we present a brief review of some recent results obtained in Rational Extended Thermodynamics, suggesting that this theory could be useful in modeling granular materials.

1 Introduction

Rapid granular flows are frequently described at the macroscopic level by the fluid-dynamic equations, suitably modified in order to take into account the dissipation due to the collisions among particles. This is the *continuum approach*, see for example the papers by Haff [1] and Capriz [2–4]. The limit of this method is that the transport coefficients are unknown and also it is not clear the validity limit of the theory. Alternative to the continuum approach is the *kinetic approach*, which makes use of the method borrowed by the theory of rarefied gases. In this framework the mean free path of the grains are supposed much larger than the typical particle size (see, for example, Cercignani et al. [5], Benedetto et al. [6], Bobylev et al. [7]).

The prototype of Extended Thermodynamics (ET) [8] is the Grad 13-moment theory [9] and several authors have used the methods of Grad to attack the problem of granular materials (see, for example, Jenkins and Richman [10] and Bisi et al. [11]).

On the other hand, in the case of rarefied gases, we know that in some situations the results of Grad are unsatisfactory when compared to the experiments. Examples concern sound waves in the limit of high frequencies, light scattering and shock waves [8]. In all these cases it is necessary to increase the number of moments. In such situations ET methodology has revealed itself to be successful and now we have a well-established theory [8]. Therefore taking into account the analogy between rarefied gases and granular materials, as well as the fact that ET is *in the middle* between kinetic and macroscopic theories, perhaps the approach of ET may be useful also to obtain new results in the research field of granular materials.

For this reason we present in this chapter the fundamentals of ET along with some recent mathematical results, as a starting point for a possible new approach in the granular materials theory.

2 Boltzmann Equation and Moments

The kinetic theory describes the state of a rarefied gas through the phase density $f(\mathbf{x}, t, \mathbf{c})$ where $f(\mathbf{x}, t, \mathbf{c})d\mathbf{c}$ is the number density of atoms at the point $\mathbf{x}$ and time t that have velocities between $\mathbf{c}$ and $\mathbf{c} + d\mathbf{c}$. The phase density obeys the Boltzmann equation

$$\partial_t f + c_i \, \partial_i f = Q, \tag{1}$$

in which the right-hand side is due to collisions between the atoms. It is well known that macroscopic thermodynamic quantities are identified as moments of the phase density

$$F_{k_1 k_2 \cdots k_j} = \int f c_{k_1} c_{k_2} \cdots c_{k_j} \, dc, \tag{2}$$

and the moments satisfy a hierarchy of balance laws in which the flux in one equation becomes the density in the next one:

$$\partial_t F + \partial_i F_i = 0$$
$$\swarrow$$
$$\partial_t F_{k_1} + \partial_i F_{ik_1} = 0$$
$$\swarrow$$
$$\partial_t F_{k_1 k_2} + \partial_i F_{ik_1 k_2} = P_{k_1 k_2}$$
$$\swarrow$$
$$\partial_t F_{k_1 k_2 k_3} + \partial_i F_{ik_1 k_2 k_3} = P_{k_1 k_2 k_3}$$
$$\vdots$$
$$\partial_t F_{k_1 k_2 \ldots k_n} + \partial_i F_{ik_1 k_2 \ldots k_n} = P_{k_1 k_2 \ldots k_n}$$
$$\vdots$$

Taking into account that $P_{kk} = 0$, the first five equations are conservation laws and coincide (using different symbols) with the well known conservation of mass, momentum and energy, respectively.

If we chose instead

$$h = k \int f \log f \, dc, \quad h^i = k \int f \log f \, c^i \, dc \tag{3}$$

as momentum, we obtain

$$\partial_t h + \partial_i h^i = g \leq 0 \tag{4}$$

representing the balance of entropy when we identify $-h$, $-h^i$ and $-g$ as the entropy density, the entropy flux and the entropy production respectively.

2.1 The Closure of Extended Thermodynamics

When we stop the hierarchy at the density with tensor of rank n, we have the problem of closure because the last flux and the productions terms are not in the list of the densities. The idea of Rational Extended Thermodynamics [8] is to consider the truncated system a as phenomenological system of continuum mechanics and then we consider the new quantities as constitutive functions

$$F_{k_1 k_2 \ldots k_n k_{n+1}} \equiv F_{k_1 k_2 \ldots k_n k_{n+1}} (F, F_{k_1}, F_{k_1 k_2}, \ldots F_{k_1 k_2 \ldots k_n})$$
$$P_{k_1 k_2 \ldots k_j} \equiv P_{k_1 k_2 \ldots k_j} (F, F_{k_1}, F_{k_1 k_2}, \ldots F_{k_1 k_2 \ldots k_n}) \quad 2 \leq j \leq n$$

to be determined.

According with the continuum theory, the restrictions on the constitutive equations come only from *universal principles*, i.e.: *the entropy principle, the objectivity Principle* and *Causality and Stability* (convexity of the entropy). We discuss later the implications of these three principles.

2.2 Macroscopic Approach of ET in the 13 Fields

The first attempt of ET was the 13-moments case. Thirteen is a special number because the first thirteen moments have a physical meaning:

$$\partial_t F + \partial_i F_i = 0$$
$$\partial_t F_{k_1} + \partial_i F_{i k_1} = 0$$
$$\partial_t F_{k_1 k_2} + \partial_i F_{i k_1 k_2} = P_{k_1 k_2}$$
$$\partial_t F_{k_1 kk} + \partial_i F_{i k_1 kk} = P_{k_1 kk}$$

The constitutive quantities are in the present case : $F_{i<k_1 k_2>}$, $F_{i k_1 kk}$, $P_{<k_1 k_2>}$ and $P_{k_1 kk}$ (the $<>$ indicates the deviatoric – traceless – part of the tensor).

The restrictions of ET due to the universal principles – in particular the entropy principle – of ET are so strong that, at least for processes not too far from equilibrium, the system is completely closed and the results are in perfect agreement with the kinetic closure procedure proposed by Grad [9]. In this case the closed system becomes (employing the usual symbols):

$$\frac{\partial}{\partial t}\rho + \frac{\partial}{\partial x^i}(\rho v_i) = 0$$

$$\frac{\partial}{\partial t}(\rho v_j) + \frac{\partial}{\partial x^i}(\rho v_i v_j + p\delta_{ij} - \sigma_{ij}) = 0$$

$$\frac{\partial}{\partial t}\left(\rho e + \rho\frac{v^2}{2}\right) + \frac{\partial}{\partial x^k}\left\{\left(\rho e + \rho\frac{v^2}{2} + p\right)v_k + q_k - \sigma_{kj}v_j\right\} = 0 \qquad (5)$$

$$\frac{\partial}{\partial t}\left\{\rho\left(v_i v_j - \frac{v^2}{3}\delta_{ij}\right) - \sigma_{ij}\right\} + \frac{\partial F_{<ij>k}}{\partial x_k} = \tau_o\,\sigma_{ij}$$

$$\frac{\partial}{\partial t}\left\{(\rho v^2 + 5p)\,v_k + 2q_k - 2\sigma_{kj}v_j\right\} + \frac{\partial F_{ppik}}{\partial x_k} = 2\tau_o\sigma_{kj}v_j - \tau_1 q_k$$

where

$$F_{<ij>k} = F_{ijk} - \frac{1}{3}F_{hhk}\delta_{ij};$$

$$F_{ijk} = \rho v_i v_j v_k + \left(pv_k + \frac{2}{5}q_k\right)\delta_{ij} + \left(pv_i + \frac{2}{5}q_i\right)\delta_{jk} + \left(pv_j + \frac{2}{5}q_j\right)\delta_{ik};$$

$$F_{ppij} = \left(\rho v^2 + 7p\right)v_i v_j + (p\delta_{ij} - \sigma_{ij})v^2 -$$
$$- \sigma_{ik}v_k v_j - \sigma_{jk}v_k v_i + \frac{14}{5}(q_i v_j + q_j v_i) + \frac{4}{5}q_k v_k\delta_{ij} + \frac{p}{\rho}(5p\delta_{ij} - 7\sigma_{ij}).$$

The first five equations are the usual conservation laws of mass, momentum and energy, while the last two blocks are evolution balance laws for the shear stress σ_{ij} and for the heat flux q_i, respectively. They reduce to the Navier–Stokes and Fourier constitutive equations when some relaxation times are small [8].

We would like to stress that in the present case we have assumed the system (5) motivated by the kinetic theory but, after that, our procedure was completely macroscopic forgetting that the $F's$ are related to a distribution function. It is very interesting to observe that the macroscopic universal principles give results which are in perfect agreement with the kinetic considerations obtained by Grad.

3 Extended Thermodynamics of Moments

Unfortunately, for rarefied gases we have discovered that in limit situations as high frequencies for sound waves, or special angles for light scattering, or large Mach number in shock waves, the 13-moment theory gives better results with respect to the Fourier–Navier–Stokes one, but its predictions are unsatisfactory when the results coming from the theory are compared to experiments.

In such situations we need more moments. In this case it is too difficult to proceed with a pure macroscopic theory (as the 13-moments theory) and it is necessary to recall that the $F's$ are moments of a distribution function f. To explain this approach we first rewrite the hierarchy of balance laws in the more compact notation

$$\partial_t u^A + \partial_i F^{iA} = g^A, \tag{6}$$

with

$$u^A = \int f c^A dc; \quad F^{iA} = \int f c^i \, c^A dc; \tag{7}$$

$$g^A = \int Q c^A dc, \tag{8}$$

where c^A

$$c^A = \begin{cases} 1 & \text{for } A = 0 \\ c_{i_1} c_{i_2} \cdots c_{i_A} & \text{for } 1 \le A \le n \end{cases}$$

and

$$u^A = \begin{cases} u & \text{for } A = 0 \\ u_{i_1 i_2 \cdots i_A} & \text{for } 1 \le A \le n \end{cases}; \quad F^{iA} = \begin{cases} u_i & \text{for } A = 0 \\ u_{i i_1 i_2 \cdots i_A} & \text{for } 1 \le A \le n \end{cases}$$

the indices i and $i_1 \le i_2 \le \cdots \le i_A$ assume the values $1, 2, 3$.

We require for the truncated system (6) the compatibility with an entropy law, i.e. all the solutions of (6) must satisfy also the supplementary entropy balance law (4) where h, h^i and g are functionals of f trough the moments (6) with $A = 0, \ldots, n$.

This condition becomes now a strong restriction for the distribution function f and the problem is: For which distribution function f_n all the classical solutions of (6) with (7) and (8) are solutions of (4)?

It was proved by Boillat and Ruggeri [12] that the function f_n depends on $(\mathbf{x}, t, \mathbf{c})$ only trough a single variable

$$f_n \equiv f_n(\chi_n),$$

where

$$\chi_n = \sum_{A=0}^{n} u'_A(\mathbf{x}, t) c^A$$

which is a polynomial in $\mathbf{c}$. In this case we have

$$h = \int \left(\chi_n F'(\chi_n) - F(\chi_n) \right) dc; \tag{9}$$

$$h^i = \int c^i (\chi_n F'(\chi_n) - F(\chi_n)) \, dc;$$

$$g = \int Q \chi_n \, dc,$$

where the partition function $F(\chi_n)$ satisfies the relation

$$F' = \frac{dF}{d\chi_n} = f_n(\chi_n).$$

Introducing the potentials

$$h' = \sum_{A=0}^{n} u^A u'_A - h; \qquad h'_i = \sum_{A=0}^{n} F_i^A u'_A - h_i;$$

we obtain

$$h' = \int F(\chi_n)\, dc, \quad h'_i = \int F(\chi_n)c_i\, dc,$$

and, choosing the main filed variable u'_A as field, it is possible to verify that

$$u^A = \frac{\partial h'}{\partial u'_A}; \qquad F^{iA} = \frac{\partial h'_i}{\partial u'_A};$$

and the original moment system becomes closed in the main field components and symmetric hyperbolic [12]

$$H^{AB}\partial_t u'_B + H_i^{AB}\partial_i u'_B = g^A(u'_C), \tag{10}$$

where

$$H^{AB}(u'_C) = \frac{\partial^2 h'}{\partial u'_A \partial u'_B} = \int F''(\chi_n)c^A c^B dc;$$

$$H_i^{AB}(u'_C) = \frac{\partial^2 h'_i}{\partial u'_A \partial u'_B} = \int F''(\chi_n)c^i\, c^A c^B\, dc.$$

Indeed, the matrix H is positive definite provided that $F''(\chi_n) > 0$ holds, since

$$H^{AB}X_A X_B = \int F''\left(c^A X_A\right)^2 dc > 0 \quad \forall X \neq 0.$$

If we require now that $-h$ is the usual entropy density for non-degenerate gases, viz.

$$h = k \int f_n \ln f_n\, dc$$

we obtain from (9)

$$(\chi_n - k \ln F')\, F' - F = 0.$$

and by differentiation we obtain

$$f_n(\chi_n) = e^{-1+\chi_n/k}. \tag{11}$$

At the equilibrium, (11) reduces to the Maxwellian distribution function. We observe that f_n is not solution of the Boltzmann equation but we have the conjecture (open problem) that for $n \to \infty$, f_n tends to the solution of the Boltzmann equation. There are several mathematical difficulties in the full non linear case, in particular about the convergence of the momentum with the approximate distribution function (11), but the convergence is ensured at least near to the equilibrium state (see, e.g., [13–15]).

4 Maximization of Entropy

There is an alternative to Extended Thermodynamics of moments for the determination of the phase density f_n. This alternative is the maximization of entropy under constraints. The two methods are equivalent.

The maximization of entropy is a method often used in statistical mechanics for the calculation of the phase density, and over the years it has acquired a certain plausibility so that its logic seems convincing. Therefore it is important to prove the consistency – even equivalence – of Extended Thermodynamics with the maximization method.

We treat the more general case in which h is a generic functional of f

$$h = \int \psi(f) d\mathbf{c}.$$

We ask the phase density to provide a maximum of h under the constraints of fixed values for the moments u_A. With the Lagrange multipliers λ^A we form the expression

$$\int \psi(f) d\mathbf{c} + \sum_{A=0}^{n} \lambda^A \left(u_A - \int c_A f d\mathbf{c} \right) \tag{12}$$

and obtain

$$\int \left(\frac{d\psi}{df} - \sum_{A=0}^{n} \lambda^A c_A \right) \delta f d\mathbf{c} = 0.$$

Thus we have

$$\frac{d\psi}{df} = \sum_{A=0}^{n} \lambda^A c_A$$

as a necessary condition for an extremum. Hence it follows that f_n is a function of

$$\chi = \sum_{A=0}^{n} \lambda^A c_A$$

and that $\psi(f)$ has the form

$$\psi(f) = \chi_n f_n - \int f_n d\chi_n. \tag{13}$$

As a sufficient condition for a maximum, we have a restriction on the function $\psi(f)$, viz. $d^2\psi/df^2 < 0$.

Insertion of (13) into (12) gives exactly the same result as the entropy principle, since $F' = f_n$ and $F = \int f_n d\chi_n$ hold. Thus we conclude that *the maximization of entropy leads to the same result as Extended Thermodynamics of moments* [12].

The reverse is also true, since $\int (\chi_n F' - F)d\mathbf{c}$ with $F'' < 0$ provides a maximum of h under the constraint. In particular, the Lagrange multipliers λ^A are identical to the main field components u'^A.

The first Author that applied the idea of maximization of entropy in Extended Thermodynamics was Dreyer [16] (see also [17]) who started from the observation of Kogan [18] that the 13-moment phase density of Grad maximizes the entropy. The procedure of maximizing entropy was introduced in information theory and physics by Jaynes [19] and it is extensively used under the name of *Maximum Entropy Principle* (see, for example, the book by Kapur [20]).

5 Maximum Characteristic Velocity in Classical Theory

The characteristic velocities λ (in the direction of propagation having unit vector $\mathbf{n} \equiv (n^i)$) of the symmetric hyperbolic system (10) are eigenvalues of $\mathbf{G}$

$$G^{AB} = H_i^{AB} n^i - \lambda H^{AB} = k \int f(\chi)(\mathbf{c} \cdot \mathbf{n} - \lambda)c^A c^B \, d\mathbf{c}$$

and in particular the wave speed for disturbances propagating in an equilibrium state are eigenvalues of

$$\int f_M (\mathbf{c} \cdot \mathbf{n} - \lambda)c^A c^B \, d\mathbf{c}, \tag{14}$$

where f_M is the Maxwellian

$$f_M = a e^{-b(c_1^2 + c_2^2 + c_3^2)}; \quad a = \frac{\rho}{\sqrt{2\pi kT}^3}; \quad b = \frac{1}{2kT}.$$

The integrals in (14) are known and therefore it is simple to evaluate the maximum eigenvalues for increasing n.

Numerical results were obtained by Weiss [21] that has remarked an increasing value of the maximum characteristic velocity for increasing number of moments N. For instance, for $N = 20$, $\lambda_{\max} = 1.8$ and for $N = 15,180$, $\lambda_{\max} = 9.36$, where $\lambda_{\max}$ is the maximum characteristic velocity evaluated in equilibrium in sound wave unity. Therefore an interesting problem is: What is the limit of $\lambda_{\max}$ when $n \to \infty$?

Before giving an answer to this question, we have to recall the theory of principal subsystems.

6 Nesting Theories and Principal Subsystems

What kind of relation does exist between two closure theories with different n (a theory S_n and a theory S_m with $n > m$)?

Boillat and Ruggeri [22] have proved that S_m is a *principal subsystem* of S_n. They showed that S_m is obtained from S_n by setting $u'^\alpha = 0, (\alpha = m+1, \ldots, n)$ and neglecting the corresponding equations for $\alpha = m+1, \ldots, n$, i.e. if

$$S_n : \begin{cases} \dfrac{\partial u^a(u'^b, u'^\beta)}{\partial t} + \dfrac{\partial F_i^a(u'^b, u'^\beta)}{\partial x_i} = \Pi^a(u'^b, u'^\beta), \\[3mm] \dfrac{\partial u^\alpha(u'^b, u'^\beta)}{\partial t} + \dfrac{\partial F_i^\alpha(u'^b, u'^\beta)}{\partial x_i} = \Pi^\alpha(u'^b, u'^\beta) \\[2mm] a = 0, \ldots, m; \quad \alpha = m+1, \ldots, n. \end{cases}$$

than

$$S_m : \frac{\partial u^a(u'^b, 0)}{\partial t} + \frac{\partial F_i^a(u'^b, 0)}{\partial x_i} = \Pi^a(u'^b, 0).$$

For general principal subsystems it was proved that the so-called subcharacteristic conditions hold [22]

$$\lambda_{\min}^{(n)} \leq \lambda_{\min}^{(m)}; \qquad \lambda_{\max}^{(n)} \geq \lambda_{\max}^{(m)}; \qquad \forall n > m. \tag{15}$$

As a consequence, the maximum velocity does not decrease when the number of moments increase.

6.1 Example of 13-Moments Principal Subsystems

As an example, we present the principal subsystems of the 13-moment theory (5). In the present case, the components of the main field $\mathbf{u}'$ are

$$\mathbf{u}' \equiv (\xi, \Lambda_j, \zeta, \Lambda_{<ij>}, \Omega_k),$$

where

$$\xi = \frac{1}{\theta} \left\{ G - \frac{v^2}{2} + \frac{1}{2p} \sigma_{ij} v_i v_j - \frac{\rho}{5p^2} q_i v_i v^2 \right\};$$

$$\Lambda_i = \frac{1}{\theta} \left\{ v_i - \frac{1}{p} \sigma_{ij} v_j + \frac{\rho}{5p^2} \left(v^2 q_i + 2 q_j v_j v_i \right) \right\};$$

$$\zeta = -\frac{1}{\theta} \left\{ 1 - \frac{2\rho}{3p^2} q_k v_k \right\}; \tag{16}$$

$$\Lambda_{<ij>} = -\frac{1}{\theta} \left\{ \frac{1}{2p} \sigma_{ij} + \frac{\rho}{5p^2} \left(v_i q_j + v_j q_i - \frac{2}{3} v_k q_k \delta_{ij} \right) \right\};$$

$$\Omega_i = \frac{\rho}{5\theta p^2} q_i,$$

being G the chemical potential and $\lambda_{\max} = 1.65 c_S$.

Let us consider the principal subsystems that we can obtain.

The 10-moment system is a subsystem of the 13-moment system, obtained when

$$\Omega_i = 0 \quad \rightarrow \quad q_i = 0$$

and neglecting the last block equation of (5)

$$\frac{\partial}{\partial t}\rho + \frac{\partial}{\partial x^i}(\rho v_i) = 0;$$

$$\frac{\partial}{\partial t}(\rho v_j) + \frac{\partial}{\partial x^i}(\rho v_i v_j + p\delta_{ij} - \sigma_{ij}) = 0; \tag{17}$$

$$\frac{\partial}{\partial t}\left(\rho e + \rho\frac{v^2}{2}\right) + \frac{\partial}{\partial x^k}\left\{\left(\rho e + \rho\frac{v^2}{2} + p\right)v_k - \sigma_{kj}v_j\right\} = 0$$

$$\frac{\partial}{\partial t}\left\{\rho\left(v_i v_j - \frac{v^2}{3}\delta_{ij}\right) - \sigma_{ij}\right\} + \frac{\partial F_{<ij>k}}{\partial x_k} = \tau_o\,\sigma_{ij}.$$

In this case $\lambda_{\max} = 1.34c_S$.

The equilibrium Euler system is a principal subsystem of the 13- and 10-moment systems with

$$\Omega_i = 0,\ \Lambda_{<i,j>} = 0 \quad \rightarrow \quad q_i = 0,\ \sigma_{ij} = 0$$

and neglecting the last block equation of (17)

$$\frac{\partial}{\partial t}\rho + \frac{\partial}{\partial x^i}(\rho v_i) = 0;$$

$$\frac{\partial}{\partial t}(\rho v_j) + \frac{\partial}{\partial x^i}(\rho v_i v_j + p\delta_{ij}) = 0;$$

$$\frac{\partial}{\partial t}\left(\rho e + \rho\frac{v^2}{2}\right) + \frac{\partial}{\partial x^k}\left\{\left(\rho e + \rho\frac{v^2}{2} + p\right)v_k\right\} = 0.$$

The maximum velocity is now $\lambda_{\max} = 1\,c_S$.

6.2 Lower Bound Estimate and Characteristic Velocities for Large n

In the previous example we have seen the validity of the subcharacteristic condition (15), now we are able to prove the behavior of $\lambda_{\max}$ when $n \to \infty$. The $(k+1)(k+2)/2$ components of order k of the main field

$$u'_{i_1 i_2 \ldots i_k}, \qquad i_1 \le i_2 \le \cdots \le i_k,$$

can be mapped in the corresponding variables

$$u'_{pqr}, \qquad p + q + r = k,$$

where p, q, r are, respectively, the number of indices equal to $1, 2, 3$. With this notation

$$\chi_n = \sum_{p,q,r} u'_{pqr} c_1^p c_2^q c_3^r; \qquad 0 \le p + q + r \le n.$$

Theorem 1 (Boillat and Ruggeri [12]). *For any n we have the lower bound conditions*

$$\frac{\lambda_{\max}}{c_S} \ge \sqrt{\frac{6}{5}\left(n - \frac{1}{2}\right)}, \tag{18}$$

where c_S is the sound velocity. Therefore $\lambda_{\max}$ becomes unbounded when $n \to \infty$.

Sketch of the proof: Using the variable u'_{pqr}, the components of the matrix $\mathbf{G}$ are given by

$$n^i \frac{\partial^2 h'_i}{\partial u'_{pqr} \partial u'_{stu}} - \lambda \frac{\partial^2 h'}{\partial u'_{pqr} \partial u'_{stu}} = \int f_M (c_i n^i - \lambda) c_1^{p+s} c_2^{q+t} c_3^{r+u} dc.$$

We know that $\mathbf{H}$ is positive definite, while $\mathbf{G}$ is semi-definite negative, if λ is the largest eigenvalue $\lambda_{\max}$. Now the elements a_{ij} of a semi-definite matrix satisfy the inequalities

$$a_{ii} a_{jj} \ge a_{ij}^2$$

and therefore we must have

$$\int f_M(c_i n^i - \lambda_{\max}) c_1^{2p} c_2^{2q} c_3^{2r} dc$$

$$\int f_M(c_i n^i - \lambda_{\max}) c_1^{2s} c_2^{2t} c_3^{2u} dc \ge \tag{19}$$

$$\left(\int f_M(c_i n^i - \lambda_{\max}) c_1^{p+s} c_2^{q+t} c_3^{r+u} dc \right)^2.$$

In this case (19) reduces to

$$\lambda_{\max}^2 \int f_M c_1^{2p} c_2^{2q} c_3^{2r} dc \int f_M c_1^{2s} c_2^{2t} c_3^{2u} dc \ge$$

$$\left(\int f_M(c_i n^i - \lambda_{\max}) c_1^{p+s} c_2^{q+t} c_3^{r+u} dc \right)^2. \tag{20}$$

With the choice $p = n, s = n - 1, q = r = t = u = 0, \mathbf{n} \equiv (1, 0, 0)$, this inequality becomes

$$\lambda_{\max}^2 \ge \frac{\int f_M c_1^{2n} dc_1}{\int f_M c_1^{2(n-1)} dc_1} =$$

$$= \frac{1}{b} \frac{\Gamma(n + 1/2)}{\Gamma(n - 1/2)} = \frac{6}{5} c_S^2 \left(n - \frac{1}{2} \right)$$

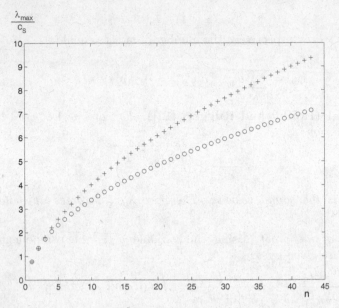

Fig. 1. The behavior of the maximum characteristic velocity versus the truncation number n and the lower bound estimate (18)

and the proof is complete. Therefore

$$\lim_{n\to\infty} \lambda_{\max} = \infty.$$

In Fig. 1 we compare the numerical values of $\lambda_{\max}/c_S$ given by Weiss [21] with the right-hand side of our lower bound (18).

This is a very surprising result because the first motivation of ET was to repair the paradox of infinite velocity of the Fourier–Navier–Stokes classical approach. Therefore for any finite n we have symmetric hyperbolic systems with finite characteristic velocities but when we take infinite moments we have a parabolic behavior.

Instead, in relativistic context it was proved that the limit of the maximum characteristic velocity for $n \to \infty$ is the light velocity [23, 24].

7 Qualitative Analysis

The ET equations are a particular case of a system of balance laws

$$\partial_t \mathbf{u} + \partial_x \mathbf{F}(\mathbf{u}) = \mathbf{f}(\mathbf{u}), \tag{21}$$

where $\mathbf{u}, \mathbf{F}$ and $\mathbf{f}$ are $\mathbb{R}^N$ vectors.

The production term $\mathbf{f}(\mathbf{u})$ represents the dissipation but, unfortunately, as we have seen, not all the components of $\mathbf{f}$ are different from zero

$$\mathbf{f}(\mathbf{u}) \equiv \begin{pmatrix} 0 \\ \mathbf{g}(\mathbf{u}) \end{pmatrix}; \qquad \mathbf{g} \in \mathbb{R}^{N-M}.$$

7.1 Shizuta–Kawashima Condition

The coupling condition discovered by Shizuta and Kawashima (K-condition) [25, 26] states that the dissipation term present in the second block of the equations has an effect also on the first block. This plays a very important role in this case for global existence of smooth solutions. The condition reads:

In the equilibrium manifold any characteristic eigenvector is not in the null space of $\nabla \mathbf{f}$, i.e.:

$$\nabla \mathbf{f} \cdot \mathbf{d}^{(i)} \Big|_E \neq 0 \qquad \forall\, i = 1, \dots, N, \tag{22}$$

where $\mathbf{d}^{(i)}$ are the right-eigenvectors of the hyperbolic system (21)

$$(\mathbf{A} - \lambda \mathbf{I})\,\mathbf{d} = 0; \qquad \mathbf{A} = \nabla \mathbf{F}$$

and E stands for the equilibrium state, i.e.:

$$\mathbf{f}(\mathbf{u}_E) = 0. \tag{23}$$

7.2 Global Existence of Smooth Solutions

If (21) is endowed with a convex entropy law, with $h(\mathbf{u})$ convex entropy function, and the system (21) is dissipative (see [22, 27] for the appropriate definition), then the K-condition becomes a sufficient condition for the existence of global smooth solutions provided that the initial data are sufficiently smooth. Hanouzet and Natalini [28] in one-space dimension and Yong [29] in the multidimensional case, have proved the following theorem:

Theorem 2 (Global existence of smooth solutions). *Assume that the system (21) is strictly dissipative and the K-condition is satisfied. Then there exists* $\delta > 0$, *such that, if* $\|\mathbf{u}(x, 0)\|_2 \leq \delta$, *there is a unique global smooth solution, which verifies*

$$\mathbf{u} \in \mathcal{C}^0\left([0, \infty);\ H^2(\mathbb{R}) \cap \mathcal{C}^1\left([0, \infty); H^1(\mathbb{R})\right)\right)$$

Moreover Ruggeri and Serre [30] have proved in the one-dimensional case that the constant states are stable:

Theorem 3 (Stability of Constant State). *Under natural hypotheses of strongly convex entropy, strict dissipativeness, genuine coupling and "zero mass" initial for the perturbation of the equilibrium variables, the constant solution stabilizes*

$$\|\mathbf{u}(t)\|_2 = O\left(t^{-1/2}\right).$$

In both theorems it plays an important role the possibility to put the system (21) in the symmetric form thanks to the introduction of the *main field* $\mathbf{u}' = \nabla h$ introduced first by Boillat [31] in a classical context and by Ruggeri and Strumia [32] in a covariant formulation.

There are many examples of dissipative systems satisfying the K-condition: the p-system with damping, the Suliciu model for the isothermal viscoelasticity, the Kerr–Debye model in non linear electromagnetism and the Jin–Xin relaxation model. Moreover quite recently it was proved that the K-condition is true also for the Extended Thermodynamics of gases [33] and for binary mixture of Euler fluids in the presence of chemical reaction [34,35].

Nevertheless the K-condition is only a sufficient condition for the global existence of smooth solution. In fact there are examples in which the K-condition is violated but the system have global smooth solutions (e.g., [36]). Recently Lou and Ruggeri [37] have observed that the K-conditions is necessary (but not sufficient) condition for genuine non linear waves.

8 Comparison with Experiments: Sound Waves and Light Scattering

The ET is very successful when the results are compared to experiments, in particular for what concerns sound waves in the limit of high frequencies and light scattering. In Fig. 2, taken from the book [8], we can see that the so-called dynamic factor $S(x, y)$ obtained by the ET fits very well the experimental data (represented in the figure by dots) when n is sufficiently large.

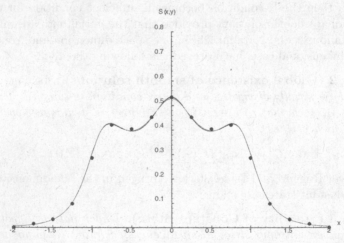

Fig. 2. Dynamical factor: the perfect agreement between the ET and the experiments

Acknowledgments

This work was supported by fondi MIUR Progetto di interesse Nazionale Problemi Matematici Non Lineari di Propagazione e Stabilità nei Modelli del Continuo (Coordinator: T. Ruggeri) and by the GNFM-INDAM.

References

1. Haff P.K.: Grain flow as a fluid-mechanical phenomenon. J. Fluid Mech., **134**, 401 (1983)
2. Capriz G.: Elementary preamble to a theory of granular gases. Rend. Sem. Mat. Univ. Padova, **110**, 179–198 (2003)
3. Capriz G., Mullenger G.: Dynamics of granular fluids. Rend. Sem. Mat. Univ. Padova, **111**, 247–264 (2004)
4. Capriz G.: Pseudofluids. In: Capriz G., Mariano P.M. (ed) Material substructures in complex bodies: from atomic level to continuum. Elsevier (2006) in print
5. Cercignani C., Illner R., Pulvirenti M.: The mathematical theory of dilute gases. Springer Series in Applied Mathematical Sciences, **106**. Springer, Berlin (1994)
6. Benedetto D., Caglioti E., Pulvirenti M.: A kinetic equation for granular media. Model. Math. Anal. Numer., **31**, 615 (1997)
7. Bobylev A.V., Carrillo J.A., Gamba I.: On some properties of kinetic and hydrodynamics equations for inelastic interactions. J. Stat. Phys., **98**, 743–773 (2000)
8. Müller I., Ruggeri T.: Rational Extended Thermodynamics. Springer Tracts in Natural Philosophy **37**, 2nd ed. Springer, New York (1998)
9. Grad H.: On the kinetic theory of rarefied gases. Comm. Pure Appl. Math., **2**, 331–407 (1949)
10. Jenkins J.T., Richman M.W.: Grad's 13–moments system for a dense gas of inelastic spheres. Arch. Ration. Mech. Anal., **87**, 355–377 (1985)
11. Bisi M., Spiga G., Toscani G.: Grad's equations and hydrodynamics for weakly inelastic granular flows. Phys. Fluids, **16**, 12, 4235–4247 (2004)
12. Boillat G., Ruggeri T.: Moment Equations in the Kinetic Theory of Gases and Wave Velocities. Cont. Mech. Thermodyn. **9**, 205–212 (1997)
13. Junk M.: Domain of definition of Levermores Five Moments System. J. Stat. Phys., **93**, 1143–1167 (1998)
14. Ruggeri T.: On the non-linear closure problem of moment equation. In: Ciancio V, Donato A, Oliveri F, Rionero S (eds) Proceedings WASCOM 99 10th Conference on Waves and Stability in Continuous Media, pp. 434–443. World Scientific (2001)
15. Brini F., Ruggeri T: Entropy Principle for the Moment Systems of degree alpha associated to the Boltzmann Equation. Critical Derivatives and Non Controllable Boundary Data. In coll. con F. Brini. Cont. Mech. Thermodyn. **14** (2), 165–189 (2002)
16. Dreyer W.: Maximization of the entropy in non-equilibrium. J. Phys. A: Math. Gen., **20** (1987)

17. Levermore C.D.: Moment Closure Hierarchies for Kinetic Theories. J. Stat. Phys. **83**, 1021 (1996)
18. Kogan M.N.: in Proc. 5th Symp. on Rarefied Gas Dynamics, **1**, suppl. 4, Academic Press, New York, 359–368 (1967)
19. Jaynes E.T.: Information Theory and Statistical Mechanics. Phys. Rev. **106**, 620 (1957)
20. Kapur J.N.: Maximum entropy models in science and engineering. John Wiley, New York (1989)
21. Weiss W.: Zur Hierarchie der Erweiterten Thermodynamik. Dissertation, TU Berlin (1990)
22. Boillat G., Ruggeri T.: Hyperbolic principal subsystems: entropy convexity and subcharacteristic conditions. Arch. Rat. Mech. Anal., **137**, 305–320 (1997)
23. Boillat G., Ruggeri T.: Maximum Wave Velocity in the Moments System of a Relativistic Gas. In coll. con G. Boillat. Cont. Mech. Thermodyn. **11**, pp. 107–111 (1999)
24. Boillat G., Ruggeri T.: Relativistic Gas: Moment Equations and Maximum Wave Velocity. In coll. con G. Boillat. J. Math. Phys., **40**, No. 12, pp. 6399–6404 (1999)
25. Shizuta Y., Kawashima S.: Systems of equations of hyperbolic-parabolic type with applications to the discrete Boltzmann equation. Hokkaido Math. J., **14**, 249–275 (1985)
26. Kawashima S.: Large-time behavior of solutions to hyperbolic-parabolic systems of conservation laws and applications. Proc. Roy. Soc. Edimburgh, **106A**, 169 (1987)
27. Boillat G., Ruggeri T.: On the shock structure problem for hyperbolic system of balance laws and convex entropy. Continuum Mech. Thermodyn., **10**, 285 (1998)
28. Hanouzet B., Natalini R.: Global existence of smooth solutions for partially dissipative hyperbolic systems with a convex entropy. Arch. Rat. Mech. Anal., **169**, 89 (2003)
29. Yong W.A.: Entropy and global existence for hyperbolic balance laws. Arch. Rat. Mech. Anal., **172**, 247 (2004)
30. Ruggeri T., Serre D.: Stability of constant equilibrium state for dissipative balance laws system with a convex entropy. Quarterly of Applied Math., **62**, 163–179 (2004)
31. Boillat G.: Sur l'existence et la recherche déquations de conservation supplémentaires pour les Systémes Hyperboliques. C. R. Acad. Sc. Paris, **278A**, 909 (1974). CIME Course, Recent Mathematical Methods in Nonlinear Wave Propagation. In: Ruggeri T. (ed) Lecture Notes in Mathematics, **1640**, 103–152, Springer (1995)
32. Ruggeri T., Strumia A.: Main field and convex covariant density for quasi-linear hyperbolic systems. Relativistic fluid dynamics. Ann. Inst. H. Poincaré, **34A** 65–84 (1981)
33. Ruggeri T.: Global existance of smooth solutions and stability of the constant state for dissipative hyperbolic systems with applications to extended thermodynamics, in *Trends and Applications of Mathematics to Mechanics*, STAMM 2002. p. 215, Springer-Verlag (2005)
34. Ruggeri T.: Some Recent Mathematical Results in Mixtures Theory of Euler Fluids, in *Proceedings WASCOM 2003*, R. Monaco, S. Pennisi, S. Rionero and T. Ruggeri (Eds.), p. 441 World Scientific-Singapore

35. Ruggeri T.: Global Existence, Stability and Non Linear Wave Propagation in Binary Mixtures of Euler Fluids, in *New Trends in Mathematical Physics*. Convegno in onore di S. Rionero, Napoli 2003. p. 205 World Scientific (2005)
36. Zeng Y.: Gas dynamics in thermal nonequilibrium and general hyperbolic systems with relaxation, Arch. Ration. Mech. Anal. **150**, no. 3, 225 (1999)
37. Lou J., Ruggeri T.: Acceleration waves and weaker Shizuta-Kawashima condition. Rendiconti del Circolo Matematico di Palermo, Serie II, **78**, 305 (2006)

Influence of Contact Modelling
on the Macroscopic Plastic Response
of Granular Soils Under Cyclic Loading

R. García-Rojo, S. McNamara, and H.J. Herrmann

Summary. An alternative to the use of continuous equations and constitutive models is the microscopic description of the material in terms of the grains themselves and the contacts (interactions) between them. This approach has been successfully applied in recent years to the study of many different problems in soil mechanics and granular physics. An open question is how realistic the microscopic model must be in order to accurately describe the macroscopic behavior observed in experiments. The objective of this contribution is to show the influence of different simple models of compacted granular soils on the overall elasto-plastic response of the system as a whole. We will focus our investigation on granular ratcheting, which is the persistent strain accumulation that a granular soil suffers under certain cyclic stress conditions. The direct influence of different models on the ratcheting response of the material will also help us to understand further this peculiar behavior of the system. The influence of particle shape will be also discussed.

1 Introduction

The differences between continuum and discrete methods perfectly reflect the different ways of approaching problems in soil mechanics. On the one hand, a continuum description of the material is possible based on well established constitutive equations, whose parameters are usually measured experimentally. On the other hand, a discrete description will directly take into account that the material is composed of distinct grains or particles that interact with each other. The final aim of this micro-structural approach is, however, to find macroscopic state variables in terms of micro-variables such as contact forces, grain displacements, local interactions, etc., in the same way that hydrodynamic fields can be connected with motion of molecules in a fluid. There is however no analog to Kinetic Theory in soil mechanics, although some useful results are available connecting macroscopic mechanical variables with a local, microscopic description of the material [1, 2].

Molecular Dynamics algorithms (MD) has been extensively used during more than fifty years for the numerical solution of a wide variety of problems.

Although molecular fluids were the original application of the method [3], it has been also applied successfully to the study of granular materials in their diverse forms [4]. In the field of soil mechanics, the term discrete element method is often applied to this technique, in order to emphasize the differences with finite element methods. This general name should be understood as referring to all numerical solving methods in which the dynamics of the grains are solved. In this sense, the term discrete element method is general and includes the Molecular Dynamics based methods (usually called Distinct Element Methods [5]) as well as more recent algorithms in which the basic unit are also the grains and their interactions [6]. The Non Smooth Contact Dynamics (CD) is therefore also a discrete element method, in the sense that the evolution of the system is solved reproducing the dynamics of the particles in terms of their inter-particle interactions [7]. It has been profusely used in the investigation of force networks and contact forces [8–11].

The aim of this chapter is the investigation of the influence of contact modelling on the overall macroscopic response of a granular material subjected to a stress-controlled cyclic loading experiment. In such experiment and above the shakedown limit, there is a plastic response of the system characterized by a constant strain-rate and a cyclic behavior of the sliding contacts (usually called *ratcheting*) [12, 13]. In order to investigate these phenomena, we first reproduce several loading cycles and analyze the differences found using different simulation methods, namely the MD and CD algorithms. This will also lead us to the briefly discuss the response of the system to a gradual increase of pressure up to the point where deformation starts. The applicability of both MD and CD discrete schemes to the study of the micro-mechanics of our simple granular soil model will be also briefly discussed. Results from MD and CD simulations of a dense system of spheres under the conditions of a biaxial test will be presented.

Recently, a thorough comparison of both methods has been presented for the case of a very simple granular packing [14]. In that reference, the influence of the iterative process inherent to CD, on the indeterminacy of the method has also been discussed and has been shown to be relevant for the overall material response. In this chapter, we want to investigate the relevance of the contact law used in the MD algorithm. For that purpose, results of the simulation of a granular packing using a linear and a non-linear contact law will be discussed.

This chapter is organized as follows. In Sect. 2 some basic features of MD and CD methods are presented. In Sect. 3, we present the results of the simulation of rigid and deformable particles under biaxial test conditions. We use Contact Dynamics for rigid particles and Molecular Dynamics for deformable ones. In this latter case, we show also results of two different contact laws. We conclude, in Sect. 4, with the discussion of the results.

2 Discrete Element Methods

A granular medium is a physical system composed by distinct basic units (grains) of a macroscopic size (typically bigger that $1\,\mu m$). If the material is dry and non-cohesive, the only interactions between grains are friction and repulsion. In this chapter, we will stick to the usual case of Mohr–Coulomb friction. The discrete character of the medium results in a complex behavior during loading and unloading that cannot be described properly up to now by any constitutive equation. Given the nature of the system, it is possible, however, to solve numerically the evolution of the grains once a valid model has been established.

Discrete element methods fit by construction and nature in a microscopic description of the granular medium. In this approach, the nature of the *grains*, and their interactions, fully determine the material response. How these interactions are modeled is an interesting subject itself, since contact modelling is a key point for any discrete model [15]. Currently used discrete methods can be basically divided into two main categories, depending on the nature of the particles (soft or infinitely hard). The simplest model reproducing most of the key features of granular material is an assembly of disks (or spheres, in 3D). In the MD method, the disks are soft in the sense that they can overlap, and the interaction between them is visco-elastic and proportional to the overlap. This idealization mimics the deformation that two real grains experience in their collision (see Fig. 1). The dynamics of the system is then solved in fixed time steps in which disturbances propagate only to the closest neighbors. In CD, the grains are rigid and an implicit iterative algorithm is used, which has the significant advantage that the implementation of friction is straightforward.

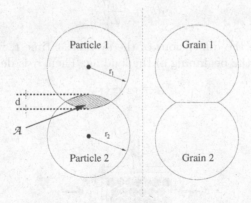

Fig. 1. Model of the deformation of the grains and contact forces. The particles (idealized grains on the *right* side of the figure) are allowed to overlap, but they are subjected then to an elastic force proportional to the overlapping that pull them apart. The interaction of two grains is represented on the *left* side of the figure, where their deformation is explicitly shown

We will describe the algorithms used for the MD and CD simulations presented in this chapter, after dealing with the boundary conditions used in our simulations.

2.1 Boundary Conditions: Biaxial Test

The biaxial test is often used in engineering to characterize the stress–strain behavior of materials. A sample is closed in a rectangular test chamber, and subjected to a confining pressure. Then a force is applied to the fully mobile box walls, so that $\sigma_1 \neq \sigma_2$, as shown in Fig. 2. In our experiments, we put a granular material in the biaxial test chamber, and start with $\sigma_1 = \sigma_2$. The walls compress the originally dilute material into a dense packing. Then, σ_2 increases gradually until the sample starts to deform, i.e., σ_1 and σ_2 obey

$$\sigma_1 = P_0, \ \sigma_2 = P_0(1 + \Delta\sigma f(t)), \tag{1}$$

where $f(t)$ is a function of time, and the $\Delta\sigma$ controls the loading intensity. Note that $\Delta\sigma$ is basically the maximum value of the deviatoric stress reached during the test. We choose $f(t)$ so that the system is loaded quasi-statically, that is, it passes through a series of stationary states, up to the value of $\Delta\sigma f(t_0)$ where the sample yields. The response of the system will be measured by the dimensional quantity γ, which is be defined in terms of the deviatoric permanent strain. This, analogously to the deviatoric stress, is the difference of the strains in the principal directions. Let the permanent strains in the principal directions be

$$\epsilon_1(t) = \frac{L_x(t)}{L_x^0}, \tag{2}$$

$$\epsilon_2(t) = \frac{L_y(t)}{L_y^0}, \tag{3}$$

where $L_{x/y}(t)$ are the dimensions of the system at time t, whereas $L_{x/y}^0$ are the dimension at the beginning of the loading. Then, γ is defined as

$$\gamma = \epsilon_2 - \epsilon_1. \tag{4}$$

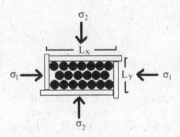

Fig. 2. Hambly's principle for biaxial test. The degrees of freedom of the walls allows to impose any pair of stresses σ_1, σ_2 to the system

Sometimes it is interesting to define the recoverable resilient strain, γ_R, accumulated along a cycle. Similarly to (4), the resilient deviatoric strain is defined in terms of the resilient strains as

$$\gamma^R = \epsilon_2^R - \epsilon_1^R. \tag{5}$$

The definition of ϵ_1^R and ϵ_2^R are similar to that in (2) and (3). They are both measured at the final stage of the loading, just before unloading starts. More detailed information about γ_R and other resilient parameters can be found under the reference [16].

2.2 Molecular Dynamics

In MD, the contact forces are calculated by considering that the overlap between two touching particles represents the deformation that generates the collision. The most successfully applied model for the contact between two spherical bodies is the well known Hertz–Mindlin model [17], in which the contact forces in the normal (f_n) and tangential (f_t) directions are proportional to the overlap distance δ

$$f_n = -K_n \delta_n, \tag{6}$$
$$f_t = -K_t \delta_t, \tag{7}$$

with the relatively complicated expression for the contact stiffnesses

$$K_n = \frac{4}{3} E^* \sqrt{R^*} \delta_n^{1/2}, \tag{8}$$
$$K_n = 8G^* \sqrt{R^*} \delta_n^{1/2}, \tag{9}$$

being E^* is the equivalent Young modulus, G^* the equivalent shear modulus and R^* the equivalent radius of the particles. An incremental approach for the force calculation, based in the previous calculation, allow to account for dissipative micro-slip effects

$$\Delta f_t = -K_t \Delta \delta_t, \tag{10}$$

where
$$K_t(\delta_n, \delta_t, E^*, G^*, R^*, \mu, \ldots) \tag{11}$$

The periods in the previous function indicate further dependencies that might also be important (such us the stress path, or any other additional relevant variables). This law is sometimes known as Hertz–Mindlin–Deresiewicz incremental contact law.

2.3 The Normal-Dashpot Model

A simpler approach, originally proposed by Cundall and Strack [18], is easily applicable to larger systems of granular materials will be used. Let us suppose that two particles i and j first touch at time t_*. Two imaginary springs are then created, one pointing along the normal direction, and the other along the tangential direction. The two springs have different properties to account for the difference between normal and tangential forces. The normal spring simply oppose to further overlapping (as shown in Fig. 1). The constant of the spring, k_n, controls the stiffness of the contact (i.e., the typical depth of the overlapping). Besides this elastic force exert in each contact, a viscous damping is also imposed, assuring some dissipation during the collision. Thus the normal force is

$$R = k_n \delta_n - \gamma \dot{\delta}_n, \tag{12}$$

where δ_n is the length of the normal spring.

The calculation of the tangential forces T is slightly more complicated, because they must obey the Coulomb condition $|T| \leq \mu R$, where μ is the static friction coefficient. We suppose that a tangential spring is created at time t_* and it is related to the tangential force via a second spring constant k_t and a damping constant γ_t. One must first calculate a candidate tangential force

$$\tilde{T} = k_t \delta_t - \gamma \dot{\delta}_t, \tag{13}$$

where δ_t is the length of the tangential spring. Then, the Coulomb condition is enforced

$$T = \begin{cases} \tilde{T}, & |\tilde{T}| \leq \mu R, \\ sgn(\tilde{T})\mu R, & \text{otherwise.} \end{cases}$$

Several modifications of (12) and (13) are found in the literature, and used to capture more realistic or specific features. A modified version of the spring-dashpot model coupling fluid flow and particle interaction has been recently used by Olivera and Rothenburg [19] to study the effect of friction on the undrained response. These authors have that friction provides additional particle instability, whereas the macroscopic strength is significantly enhanced by increasing the friction coefficient. Other modifications of the model allow for the introduction of cohesive interactions in the sample [20].

The Normal Spring Length

An interesting point, is how the definition of the normal spring length δ_n affects the spring-dashpot model and, more specifically, the material response. The most usually variant, defines δ_n as the penetration distance between two overlapping disks (distance d in Fig. 1)

$$R = k_n d - \gamma \dot{d}. \tag{14}$$

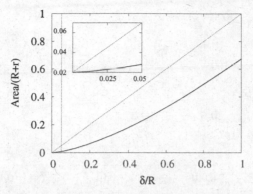

Fig. 3. Values of $\delta_n^{area}/\delta_n^{lin}$ in terms of the overlapping distance d. The typical values of d found in our simulations are shown in the inset. This ratio gives a clear idea of the deviation of the non-linear law respect to d

Another possibility is to define the spring length proportional to the overlapping area (A, in the figure). In order to be consistent with the dimensions, the normal contact law between to interacting grains with respective radii r_1 and r_2 will be in this case

$$R = k_n \frac{A}{r_1 + r_2} - \gamma \dot{p}. \tag{15}$$

Note that the repulsive force depicted in this equation is not linear with the overlap distance.

In the range of overlapping distances involved in our simulations, the behavior of the contact laws is obviously different (Fig. 3).

From the contact forces calculated with the non-linear law (15), it is possible to estimate an *equivalent* linear normal stiffness, as

$$k_l^* = \left| \frac{\Delta F_n}{\Delta p} \right|, \tag{16}$$

where Δ represents the increment of the variables. So defined, this stiffness is similar to the Hertz–Mindlin–Deresiewicz incremental contact law of (10). It is now interesting to know if this stiffness so defined is still independent of the overlapping distance d. Figure 4 shows that there is a power-law dependence, $k_l^* \propto \delta^{0.5}$. Note that this result is consistent with the dependence of the overlapping Area A with d.

2.4 Contact Dynamics

In Contact Dynamics, there is no overlapping of disks, for they are considered as perfectly rigid and interacting with each other only at the contact point. The algorithm is basically an iterative procedure after which a force network is calculated down to some precision satisfying certain physical restrictions [21]. Each contact force depends on the adjacent contact forces, which means that

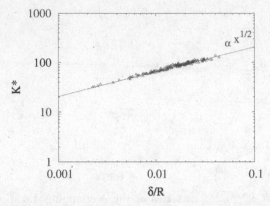

Fig. 4. Equivalent linear stiffness k_i^* as a function of the overlapping distance at the contacts in the simulation with a non-linear law in the normal spring. The overlapping distance is scaled with the mean radius of the particle R

the problem cannot be solved locally for each contact. The main constrain to be fulfilled is impenetrability. The normal force is chosen to be the smallest value R needed to avoid interpenetration. The proper tangential force is chosen that prevents the contact from sliding. If Coulomb's condition cannot be satisfied, the contact will slide with $T = \mu R$ against the relative velocity. The main drawback of CD is the indeterminacy of the forces. The important question arises: what makes the CD method choose one of these possible solution among the others? or, is the selected solution somehow special among the other admissible, or are they all equivalent [22, 23].

We conclude by noting one important difference between the CD and MD approach. In CD, the granular packing has access to any possible force network, whereas in MD, its choice is restricted, because the forces can only be modified by small motions of the grains.

3 Results

In this section we will present the results of our simulations. The system we are dealing with is composed of 100 disks with a Gaussian distribution of radii, whose mean is $R = 1\,\text{cm}$. The initial condition was obtained by compressing a random distribution of the grains up to a certain pressure P_0. The compression was carried out without friction between the particles, in order to increase the volume fraction. The same initial condition was used for all the experiments shown.

The main parameter of the MD model is the normal stiffness k_n. In our simulations $k_n = 1.610^6 \ \text{N}\,\text{m}^{-1}$. The typical frequency of the spring ω (and therefore the characteristic oscillation period $t_s = \pi/\omega$), can be defined in terms of k_n

$$\omega = \sqrt{k_n/m_{ij} - \eta_0}. \tag{17}$$

In this expression, m_{ij} is the reduced mass of the particles that interact and η_0 is the damping constant, another parameter of the simulations. In terms of this latter quantity, the relaxation time is $t_r = 1/\eta_0$. In the MD simulations, this time should be much bigger than t_s, $t_r \gg t_s$. The MD time step t_{MD}, should also taken big enough, so that $t_{MD} \gg t_s$. In the cyclic loading, we choose the typical period much bigger than the oscillation time, $t_0/t_s = 10^5$. Important parameters for the simulation are the ratio of stiffnesses k_t/k_n and the static friction coefficient. In our simulations $k_t/k_n = 0.33$, and the confining pressure $P_0 = 0.001 k_n$. Since all our experiments are in the quasistatic range, we can assume that the static and dynamics friction coefficient are equal $\mu = 0.1$. For the CD simulations, the same value of the friction is used. In the iterative process, the previous force configuration at the contacts is used as first guess, as a sensible way of implementing history dependence in the simulation [14].

3.1 Comparing MD and CD

For the comparison between the algorithms, the system is first homogeneously compressed under a certain pressure $P_0 = \frac{\sigma_1 + \sigma_2}{2}$, until a compacted state is reached. This first stage is carried out with the MD algorithm. After this preparation of the sample, two different simulations (MD and CD) are run in which the axial component of the stress, σ_2, is periodically changed. We choose the simplest expression for $f(t)$ in (1)

$$f(t) = \begin{cases} t, & if \ t < t_0/2, \\ t_0/2 - t, & if \ t > t_0/2 \end{cases}$$

Figure 5 shows the strain–stress curve after one of this cycles, obtained with the MD algorithm (left) and with the CD algorithm (right). The differences in the range of values are already obvious for this first cycle. The perturbation that the loading exerts on the system is much weaker in the CD simulation, and related to the precision of the method. But note also that the energy dissipated is bigger in the MD cycle, and so is the remanent strain at the end of the process.

The higher *inertia* of the CD method is even more clearly observed in Fig. 6, where the evolution of the strain γ is plotted for both methods in an experiment in which the load is steadily increased

$$f(t) = t.$$

In the MD simulation, the system starts expanding slowly. A sudden compression is perceived $t = 2,700\,s$, after which the deformation rate seems to grow in each cycle. This leads to a collapse of the sample at the end of the simulation. In this range of values, however, the CD simulated sample seems to remain unaltered. A closer look to its behavior is presented in the inset of the figure. In the CD simulation, the response is smoother than in the MD

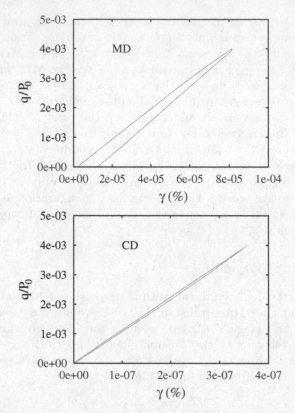

Fig. 5. Stress–strain curve obtained after the loading and unloading of a compressed sample in a MD (*top*) and CD (*below*) simulation. The initial condition was the same in both experiments. The system was compressed at a linear rate and the decompressed at the same rate until the original stress state was reached. Note the difference in horizontal axes (γ). The maximum value of the deviatoric stress in this simulation is $\Delta\sigma = 0.2$

case. At a first stage, the system seems not to be affected by the imposed loading. There is however a critical load, beyond which, the system expands. In contrast to the MD experiment, this expansion is carried out without any collapse or breakage of the physical structure of the grains.

3.2 Comparing Different Visco-Elastic Laws

The existence of granular ratcheting has been reported in MD simulations of a dense packing of polygons [12] and disks [13]. The response of a given compressed system of disks subjected to cyclic loading varies according to the imposed loading going from a resilient (elastic) response, to a regime in which the permanent accumulated deformation increases after each cycle in a fix amount.

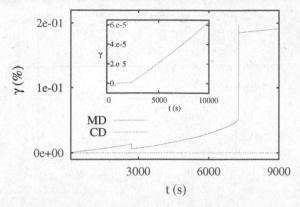

Fig. 6. Evolution of the permanent strain γ in the Molecular Dynamics and in the Contact Dynamics simulations. The inset shows, in a more appropriate scale, the behavior of γ in the CD simulation

Initial Configuration

For comparison, we have used the same initial conditions for both simulations with different contact laws. This initial configuration (a system of 100 polydisperse disks) has been therefore carefully obtained. First, the system is homogeneously compressed under a certain pressure $P_0 = \frac{\sigma_1 + \sigma_2}{2}$ using a non-linear contact law for the contacts, until a compacted state is reached.

Note that the normal spring stiffness k_n is the same for both simulations. A change in the contact law therefore implies a change in the confining pressure necessary to keep the initial configuration as well. A secondary compression is consequently necessary in the simulation with the linear law. We have calculated that the equivalent confining pressure in the linear case to be

$$P_0^{lin} = 2.2528 P_0^{area}. \tag{18}$$

Material Response

After this preparation of the sample, two different simulations are run (with a linear and a non-linear law for the normal at the contacts) in which the axial component of the stress, σ_2, is periodically changed,

$$\sigma_2(t) = P_0 \left[1 + \frac{\Delta \sigma}{2} \left(1 - \cos \left(\frac{2\pi t}{t_0} \right) \right) \right], \tag{19}$$

where t is the time and t_0 is the period of the cyclic loading. The changes in the loading are characterized by the parameter $\Delta \sigma$, which is directly related to the maximum value of the deviatoric stress.

The different contact models induces then both a microscopic and macroscopic different response of the material. On the one hand, Fig. 8 shows a

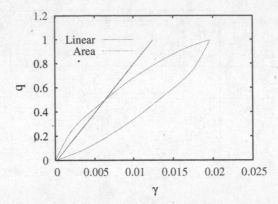

Fig. 7. Stress–strain cycles for the linear and non-linear (area) contact laws. $q = \sigma_1 - \sigma_2$ is the deviatoric strain and γ is defined in the text

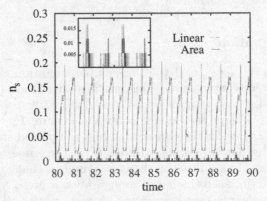

Fig. 8. The relative number of sliding contacts n_s is the ratio of sliding contacts and the total number of contacts. In the figure, n_s is plotted for both contact laws as a function of the number of cycles. In this simulation, $\Delta\sigma = 0.10$

big difference in the number of sliding contacts, although in both simulations n_s is periodical, being n_s much smaller in the linear case. This different behavior of the sliding in the system induces a diverse material behavior, reflected in the stress–strain cycles of Fig. 7, being the stiffness of the material higher in the linear case. This can be measured in terms of the resilient modulus $M_R = \Delta\sigma/\gamma_R$, which is the ratio of the maximum deviatoric stress and the corresponding deviatoric resilient strain defined in (5). M_R is noticeably higher for the simulation that uses the linear law. Observe also that in this linear case, dissipation (the area enclosed by the cycle) is almost null, when compared with the other curve in the graph.

The effect on the permanent strain accumulation is also very clear, as one can see on Fig. 9. If a linear contact law is used, the system accumulates permanent strain at a speed order of magnitude lower than the non-linear

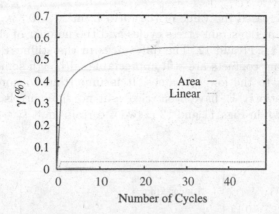

Fig. 9. Evolution of the permanent deviatoric strain γ with the number of cycles for the different contact laws

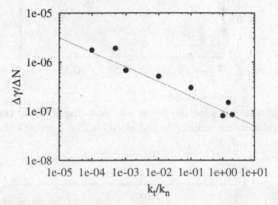

Fig. 10. Dependence of the strain-rate on the stiffness ratio k_t/k_n. Data correspond to the simulation of a system with normal damping $1/\gamma_n = 4 \times 10^3 t_s$, and tangential damping $1/\gamma_t = 8 \times 10^2 t_s$, solid fraction $\Phi = 0.845 \pm 0.005$, and friction coefficient $\mu = 0.1$. The stress conditions are kept constant, $P_0 = 10^{-3} \times k_n$ and $\Delta\sigma = 0.2$. The *solid line* represents the power law $y \propto x^{-0.3}$

case. Reason for the different behavior in the system described up to now is the change in the typical value of the ratio of the normal and tangential forces F_n/F_t that the change in the contact law implies. This is so, because we have changed F_n, but no change on the tangential law (therefore on F_t) has been done. One may think of adjusting the parameter k_n/k_t in the linear case, trying to reproduce the results of the non-linear contact law. We now know that this parameter (k_n/k_t) affects the macroscopic strain accumulation, for it is strongly related to sliding in the system [16]. Figure 10 shows that the strain rate $\Delta\gamma/\Delta N$, actually decreases as the ratio of stiffnesses k_n/k_t increases.

We have increased the value of the ratio from $k_t = 0.33k_n$ to $k_t = 4k_n$, and the results in the strain–stress cycles and the number of sliding contacts are shown in Figs. 11 and 12. The differences in the stiffness (M_R) and the number of sliding contacts are still appreciable, although some correction is observed resect to the low k_T values. It is important to note, that higher values of the ratio k_t/k_n have been tried, but no difference is observed with the results shown in Figs. 11 and 12 above a certain limit $k_t \approx 2k_t$.

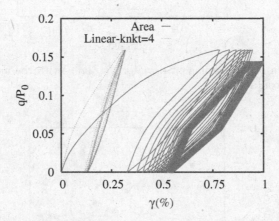

Fig. 11. Stress–strain cycles for the linear and non-linear (area) contact laws. The details of the simulation are similar to the ones on Fig. 7, except that $k_t/k_n = 4$ in this case

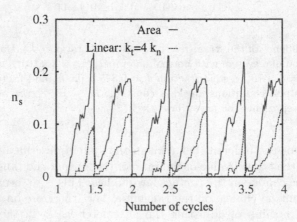

Fig. 12. Evolution of the number of sliding contacts n_s with the number of cycles for the simulation shown in Fig. 11

4 Conclusions

A different behavior of the plastic response of a system simulated by a MD algorithm and a CD algorithm has been shown. The response of the material is more apparent in the MD simulation, while CD is much more resistant to collapse. This behavior is probably related to the indeterminacy of the forces in the CD method [22, 23]. In the Contact Dynamics method, the system has many possible force configurations that are compatible with the stress conditions. The system can jump from one force configuration to another without any movement of the particles. This is not the case in the Molecular Dynamics scheme, were the forces can only be changed by small motions of the grains.

MD results have been compared with experimental triaxial tests data and a good correlation has been found [13]. No similar validation of the CD method for the repetitive loading case has been reported yet, although CD method has been successfully applied by different authors to the description of the contact forces of a static packing [8–11]. Further investigation is therefore needed in order to delimit the exact physical implications of the divergences between methods reported here.

We have shown that the law used to model the contact between grains strongly influences the material response to cyclic loading. The use of a contact law proportional to the overlapping area is equivalent to the use of a linear law with an incremental non-linear stiffness.

For equivalent initial configuration of the contacts and preparation of the sample, a linear contact law implies less sliding contacts, less accumulation of permanent strain and also a higher stiffness of the material (as measured with the resilient modulus M_R).

The parameter k_n/k_t can be used to adjust the linear contact law, but it is not possible to reproduce the same results than in the non-linear case. A systematic comparison of the simulation results with experimental data will help to determine the best contact law to use in each experiment.

Acknowledgments

The authors want to acknowledge the EU project *Degradation and Instabilities of Geomaterials with Application to Hazard Mitigation* (DIGA) in the framework of the Human Potential Program, Research Training Networks (HPRN-CT-2002-00220).

References

1. K. Bagi, Mech. of Mat. **22**, 165 (1996).
2. P. A. Cundall, A. Drescher, and O. D. L. Strack, in *IUTAM Conference on Deformation and Failure of Granular Materials* (Delft, 1982), pp. 355–370.
3. B. J. Alder and T. E. Wainwright, J. Chem. Phys. **31**, 459 (1959).

4. H. J. Herrmann and S. Luding, Continuum Mechanics and Thermodynamics **10**, 189 (1998).
5. P. A. Cundall, in *Proc. Symp. Int. Rock Mech.* (Balkema, Nancy, 1971), vol. 2.
6. J. J. Moreau, Ann. Inst. H. Poincaré Anal. Non Lin'eaire **XXX**, 1 (1989).
7. J. J. Moreau, in *Powders & Grains 93* (Balkema, Rotterdam, 1993), p. 227.
8. F. Radjai, M. Jean, J. J. Moreau, and S. Roux, Phys. Rev. Lett. **77**, 274 (1996a).
9. F. Radjai, D. Wolf, S. Roux, M. Jean, and J. J. Moreau, in *Friction, Arching and Contact Dynamics*, edited by D. E. Wolf and P. Grassberger (World Scientific, Singapore, 1997).
10. S. Roux, in *Physics of Dry Granular Media*, edited by H. J. Herrmann, J.-P. Hovi, and S. Luding (Kluwer Academic Publishers, Dordrecht, 1998), p. 267.
11. J. H. Snoeijer, T. Vlugt, M. van Hecke, and W. van Saarloos, Phys. Rev. Lett. **92**, 054302 (2004).
12. F. Alonso-Marroquin and H. Herrmann, Phys. Rev. Lett. **92**, 054301 (2004), cond-mat/0403065.
13. R. García-Rojo and H. Herrmann, Granular matter **7**, 109–118 (2005a), cond-mat 0404176.
14. R. G.-R. S. McNamara and H. J. Herrmann, Phys. Rev. E **72**, 021304 (2005).
15. S. Luding, in *Physics of dry granular media - NATO ASI Series E350*, edited by H. J. Herrmann, J.-P. Hovi, and S. Luding (Kluwer Academic Publishers, Dordrecht, 1998), p. 285.
16. R. García-Rojo and H. Herrmann, Phys. Rev. E **72**, 041302 (2005b).
17. R. D. Mindlin, J. of Appl. Mech. **16**, 259 (1949).
18. P. A. Cundall and O. D. L. Strack, Géotechnique **29**, 47 (1979).
19. R. Olivera and L. Rothenburg, in *Powders and Grains 2005*, edited by H. H. R. García-Rojo and S. McNamara (Taylor and Francis, 2005), pp. 1223–1227.
20. O. R. Walton, in *Particulate two-phase flow*, edited by M. C. Roco (Butterworth-Heinemann, Boston, 1993), p. 884.
21. F. Radjai, L. Brendel, and S. Roux, Phys. Rev. E **54**, 861 (1996b).
22. S. McNamara and H. J. Herrmann (2004).
23. T. Unger, J. Kertész, and D. Wolf (2004), cond-mat/0403089.

Fluctuations in Granular Gases

A. Barrat, A. Puglisi, E. Trizac, P. Visco, and F. van Wijland

Summary. A driven granular material, e.g. a vibrated box full of sand, is a stationary system which may be very far from equilibrium. The standard equilibrium statistical mechanics is therefore inadequate to describe fluctuations in such a system. Here we present numerical and analytical results concerning energy and injected power fluctuations. In the first part we explain how the study of the probability density function (pdf) of the fluctuations of total energy is related to the characterization of velocity correlations. Two different regimes are addressed: the gas driven at the boundaries and the homogeneously driven gas. In a granular gas, due to non-Gaussianity of the velocity pdf or lack of homogeneity in hydrodynamics profiles, even in the absence of velocity correlations, the fluctuations of total energy are non-trivial and may lead to erroneous conclusions about the role of correlations. In the second part of the chapter we take into consideration the fluctuations of injected power in driven granular gas models. Recently, real and numerical experiments have been interpreted as evidence that the fluctuations of power injection seem to satisfy the Gallavotti–Cohen Fluctuation Relation. We will discuss an alternative interpretation of such results which invalidates the Gallavotti–Cohen symmetry. Moreover, starting from the Liouville equation and using techniques from large deviation theory, the general validity of a Fluctuation Relation for power injection in driven granular gases is questioned. Finally a functional is defined using the Lebowitz–Spohn approach for Markov processes applied to the linear inelastic Boltzmann equation relevant to describe the motion of a tracer particle. Such a functional results to be different from injected power and to satisfy a Fluctuation Relation.

1 Introduction

In equilibrium thermodynamics one characterizes the stable phases of a system using a limited set of macroscopic state variables, therefore bypassing much of the microscopic details of the systems under study. It is only very recently that the same strategy has been applied to systems in Non-Equilibrium

Steady-States (NESS). And the need for such an approach is all the more pregnant for the study of NESS that no general formalism parallel to the standard equilibrium Gibbs–Boltzmann ensemble theory exists. This field started with experiments carried out on turbulent flows or convection cells, and much more recently on granular systems. Global observables, namely spatially integrated over the whole system, and their distribution, may indeed coarse-grain the irrelevant microscopic details specific to the system at hand, while allowing for comparisons between different systems. They are expected to be more robust and more exportable tools for analysis than local probes, like, e.g. structure factors. This has led to the observation of intriguing similarities between turbulent flows and granular systems [BHP98, BdSMRM05, Ber05]. However, one must take into account the key ingredient making a NESS way different from its equilibrium counterpart: steady flows of energy, matter, or else, run across the system. The existence of currents characterizes a NESS, and makes it different from an equilibrium state in that detailed balance (time reversibility) no longer holds. Given that the time direction plays a central rôle, one is led to the idea that time integration may also be useful in smoothening out various details of the microscopic dynamics. This has motivated several authors to consider the distribution of time integrated and spatially averaged quantities characterizing the NESS as such, like that of the injected power in a turbulent flow or in a granular gas.

We briefly turn to a reminder of phenomenological thermodynamics of nonequilibrium systems, as presented in [dGM69]. There, for systems only slightly away from equilibrium, the concept of entropy can be extended in a consistent fashion, and its time evolution goes according to

$$\frac{dS}{dt} = \int_V \sigma_{\text{irr}} - \int_V \nabla \cdot \mathbf{J}_S. \tag{1}$$

The intrinsic entropy production rate σ_{irr} is positive definite, and cancels under the condition that the system reaches equilibrium. The other piece in the rhs of (1), which features an entropy current $\mathbf{J}_S$, conveys the existence of external sources, often located at the system's boundaries, driving the system out of equilibrium. The entropy current is not but a linear combination of the various currents flowing through the system, with the conjugate affinities (like a temperature or a chemical potential gradient) as the proportionality factors. The entropy current – when it can univoquely be defined – therefore stands as a relevant measure of how far the system is from equilibrium. For that reason, various studies, starting from the pioneering work of Evans et al. [ECM93], in a study of a thermostatted fluid under shear, have been focused on the appropriately generalized expression of the latter entropy current. In Evans, Cohen and Morriss' case it is simply proportional to the power provided by the thermostat to compensate for viscous dissipation. They went on to determine the distribution function of $Q_S(t)$, the time integrated entropy flow (or equivalently the energy provided by the thermostat), denoted by $P(Q_S, t)$. In doing so they empirically noticed a remarkable property of the pdf of Q_S, namely

$$\lim_{t\to\infty} \frac{1}{t} \log \frac{P(Q_S, t)}{P(-Q_S, t)} = q_S, \tag{2}$$

where $q_S = Q_S/t$, which is a time-intensive quantity, is the time average of J_S over $[0, t]$. This symmetry property of the pdf of Q_S was soon to be formalized into a theorem for thermostatted systems by Gallavotti and Cohen [GC95], and has since triggered a flurry of studies. The mathematical object defined by $\pi(q_S) = \lim_{t\to\infty} \frac{\log P(q_S\ t, t)}{t}$ is seen to be extending the concept of intensive free energy to a nonequilibrium setting, and will occupy much of our numerical and analytical efforts.

In the realm of nonequilibrium systems, granular gases play a central rôle as systems exhibiting a strongly irreversible microscopic dynamics due to inelastic collisions, and for these no viable definition of entropy, let alone entropy flow, is available. This has led various authors [AFMP01, FM04] to conjecture that, by analogy to thermostatted systems, the power injected into the system to maintain it in a steady-state, could satisfy a symmetry property like the one uncovered by [ECM93], and its ensuing consequences in terms of generalized fluctuation–dissipation theorems. Fortunately, a well-controlled kinetic theory-based statistical mechanics exists for dilute gaseous systems, and we shall build upon it to investigate the questions raised above.

The outline of the present review is as follows. We begin in Sect. 2 with a brief introduction to granular gases and the basics of their statistical mechanics. In Sect. 3 we analyze the distribution of the total kinetic energy of the gas, as a first choice for a global observable. In Sect. 4 and 5 we address numerically, analytically and also experimentally, the issue of interpreting the power injected into a gas in terms of entropy flow, the negative outcome of which leads us to Sect. 6. There we construct a one particle observable exhibiting the properties expected from an entropy flow, and quite notably its distribution function displays the symmetry property (2).

2 A Brief Introduction to Granular Gases

A granular gas is an assembly of macroscopic particles kept in a gaseous steady state by a constant excitation [BTE05] (the typical example can be illustrated thinking of many beads in a strongly vibrated box). The simplest way to characterize such systems is to consider N identical smooth hard spheres, losing a part of their kinetic energy after each collision. The total momentum is conserved in collisions, and only the normal component of the velocity is affected. Thus, the collision law for a couple of particles $(1, 2)$ reads:

$$\begin{cases} \mathbf{v}_1^* = \mathbf{v}_1 - \frac{1}{2}(1+\alpha)(\mathbf{v}_{12} \cdot \hat{\boldsymbol{\sigma}})\hat{\boldsymbol{\sigma}} \\ \mathbf{v}_2^* = \mathbf{v}_2 + \frac{1}{2}(1+\alpha)(\mathbf{v}_{12} \cdot \hat{\boldsymbol{\sigma}})\hat{\boldsymbol{\sigma}}, \end{cases} \tag{3}$$

where $\hat{\boldsymbol{\sigma}}$ is a unitary vector along the center of the colliding particles at contact. Here α is a constant, called the coefficient of normal restitution ($0 \leq \alpha \leq 1$, and when $\alpha = 1$ collisions are purely elastic). Without some energy injection mechanism the total energy of the gas will decrease in time, until all the particles are at rest (cooling state). However, when some energy input is provided, the system can reach a nonequilibrium stationary state. Energy injection may be supplied in several ways, which can be divided in two main categories: injection from the boundaries and homogeneous driving. In the former category energy is supplied by a boundary condition, the system hence develops spatial gradients and it is not homogeneous. The latter category refers to systems where energy injection is achieved by a homogeneous and isotropic force acting on each particle.

2.1 Boundary Driven Gases

In this section we will give a short introduction to the methods used to describe the behavior of a granular gas in which energy is injected by a boundary condition (typically a vibrating wall). This kind of system has been widely studied in the literature [GZBN97, MB97, ML98, Kum98, BRMM00, BT02], and one of its main characteristics is that the density and the temperature are not homogeneous over the system: there is a heat flux, which does not verify Fourier law. This feature is well described by kinetic theory and in good agreement with the hydrodynamic approximation, which allows an analytical calculation of the density and temperature profiles. In the dilute limit, such a system is well described by the Boltzmann equation:

$$\partial_t f(\mathbf{r}, \mathbf{v}_1, t) + \mathbf{v}_1 \cdot \boldsymbol{\nabla} f(\mathbf{r}, \mathbf{v}_1, t) = J[f|f]. \tag{4}$$

Here $J[f|f]$ is the collision integral, which takes into account the inelasticity of the particles:

$$J[f|f] = \sigma^{d-1} \int d\mathbf{v}_2 \int' d\hat{\boldsymbol{\sigma}} (\mathbf{v}_{12} \cdot \hat{\boldsymbol{\sigma}}) \left(\frac{f(\mathbf{v}_1^{**}, t) f(\mathbf{v}_2^{**}, t)}{\alpha^2} - f(\mathbf{v}_1, t) f(\mathbf{v}_2, t) \right), \tag{5}$$

where the notation $\mathbf{v}_{12}$ denotes the relative velocity between particles 1 and 2, the two stars superscript (i.e. $\mathbf{v}^{**}$) denote the precollisional velocity of a particle having velocity $\mathbf{v}$, and the primed integral is a short-hand notation meaning that the integration is performed on all angles satisfying $\mathbf{v}_{12} \cdot \hat{\boldsymbol{\sigma}} > 0$. The hydrodynamic fields are defined as the velocity moments:

$$n(\mathbf{r}, t) = \int d\mathbf{v} f(\mathbf{r}, \mathbf{v}, t), \tag{6}$$

$$n(\mathbf{r}, t)\mathbf{u}(\mathbf{r}, t) = \int d\mathbf{v} \, \mathbf{v} f(\mathbf{r}, \mathbf{v}, t), \tag{7}$$

$$\frac{d}{2} n(\mathbf{r}, t) T(\mathbf{r}, t) = \int d\mathbf{v} \frac{m}{2} (\mathbf{v} - \mathbf{u})^2 f(\mathbf{r}, \mathbf{v}, t), \tag{8}$$

and the hydrodynamic balance equations for those quantities are derived taking the velocity moments in (4). Their expression is:

$$\partial_t n + \nabla \cdot (n\mathbf{u}) = 0, \tag{9}$$

$$(\partial_t + \mathbf{u} \cdot \nabla) u_i + (mn)^{-1} \nabla_j P_{ij} = 0, \tag{10}$$

$$(\partial_t + \mathbf{u} \cdot \nabla + \zeta) T + \frac{2}{3n} (P_{ij} \nabla_j u_i + \nabla \cdot \mathbf{q}) = 0, \tag{11}$$

where the pressure tensor P_{ij}, heat flux $\mathbf{q}$, and the cooling rate ζ are defined by:

$$P_{ij}(\mathbf{r}, t) = \int d\mathbf{v} \, m(v_i - u_i)(v_j - u_j) f(\mathbf{r}, \mathbf{v}, t), \tag{12}$$

$$\mathbf{q}(\mathbf{r}, t) = \int d\mathbf{v} \frac{m}{2} (\mathbf{v} - \mathbf{u})^2 (\mathbf{v} - \mathbf{u}) f(\mathbf{r}, \mathbf{v}, t), \tag{13}$$

$$\zeta(\mathbf{r}, t) = \frac{(1 - \alpha^2) m \pi^{\frac{d-1}{2}} \sigma^{d-1}}{4 d \Gamma \left(\frac{d+3}{2}\right) n(\mathbf{r}, t) T(\mathbf{r}, t)} \int d\mathbf{v}_1 \int d\mathbf{v}_2 |v_{12}|^3 \, f(\mathbf{r}, \mathbf{v}_1, t) f(\mathbf{r}, \mathbf{v}_2, t). \tag{14}$$

Explicit analytical expressions for the above quantities have been obtained in the limit of small spatial gradients by Brey et al. [BDKS98, BC01]. Moreover for systems in the steady state without a macroscopic velocity flow the hydrodynamic equations simplify, and therefore the temperature and density profiles can be explicitly computed.

2.2 Randomly Driven Gases

We consider here a granular gas kept in a stationary state by an external homogeneous thermostat, the so called "Stochastic thermostat", which couples each particle to a white noise. Energy injection is hence achieved by means of random forces acting independently on each particle, and drives the gas into a non-equilibrium steady state. The equation of motion governing the dynamics of each particle is therefore:

$$m \frac{d\mathbf{v}_i}{dt} = \mathbf{F}_i^{\text{coll}} + \mathbf{F}_i^{\text{th}}, \tag{15}$$

where $\mathbf{F}_i^{\text{coll}}$ is the force due to collisions and $\mathbf{F}_i^{\text{th}}$ is a Gaussian white noise (i.e. $\langle F_{i\gamma}^{\text{th}}(t) F_{j\delta}^{\text{th}}(t') \rangle = 2\Gamma \delta_{ij} \delta_{\gamma\delta} \delta(t - t')$, where the subscripts i and j are used to refer to the particles, while γ and δ denote the Euclidean components of the random force). This model is one of the most studied in granular gas theory and reproduces many qualitative features of real driven inelastic gases [WM96, PO98, PLM$^+$98, vNETP99, HBB00, MSS01, PTvNE02, vNE98]. After a few collisions per particle the system attains a non-equilibrium stationary state. This state seems homogeneous. From the equations of motion it is possible to derive the homogeneous Boltzmann equation governing the evolution of the one-particle velocity distribution function [vNE98]:

$$\partial_t f(\mathbf{v}_1, t) = J[f, f] + \Gamma \Delta_{\mathbf{v}_1} f(\mathbf{v}_1), \tag{16}$$

where the Laplace operator $\Delta_{\mathbf{v}} \equiv (\partial/\partial_{\mathbf{v}})^2$ is a diffusion term in velocity space characterizing the effect of the random force, while $J[f, f]$ is the collision integral, which takes into account the inelasticity of the collisions (cf. (5)). The granular temperature of the system is defined as usual as the mean kinetic energy per degree of freedom, $T_g = \langle v^2 \rangle / d$. The stationary solution of (16) has extensively been investigated in the last years. Even if an exact solution is still missing, a general method is to look for solutions in the form of a Gaussian distribution multiplied by a series of Sonine polynomials [CC60]:

$$f_{st}(\mathbf{v}) = e^{-\frac{v^2}{2T_g}} \left(1 + \sum_{p=1}^{\infty} a_p S_p \left(\frac{v^2}{2T_g} \right) \right). \tag{17}$$

The expression of the first three Sonine polynomials is:

$$S_0(x) = 1$$

$$S_1(x) = -x + \frac{1}{2}d \tag{18}$$

$$S_2(x) = \frac{1}{2}x^2 - \frac{1}{2}(d+2)x + \frac{1}{8}d(d+2).$$

Moreover the coefficients a_p are found to be proportional to the averaged polynomial of order p:

$$a_p = A_p \left\langle S_p \left(\frac{v^2}{2T_g} \right) \right\rangle, \tag{19}$$

where A_p is a constant and the angular brackets denote average with weight f_{st}. From this observation one directly obtains that the first coefficient a_1 vanishes by definition of the temperature. A first approximation for the velocity pdf is therefore to truncate the expansion up to the second order ($p = 2$). An approximated expression for the coefficient a_2 has been found as a function of the restitution coefficient α and the dimension d [vNE98, CDPT03, MS00]. Its expression is:

$$a_2(\alpha) = \frac{16(1-\alpha)(1-2\alpha^2)}{73 + 56d - 24\alpha d - 105\alpha + 30(1-\alpha)\alpha^2}. \tag{20}$$

It must be noted that the second Sonine approximation is only valid for not too large velocities, since the tails of the pdf have been shown [vNE98] to be overpopulated with respect to the Gaussian distribution. It is known [vNE98] that in high energies $\log f(v) \sim -(v/v_c)^{3/2}$ with a threshold velocity v_c that diverges when the dimension d goes to infinity. This means that at high dimensions the distribution is almost a Gaussian, since both the tails and the a_2 contributions tend to vanish. All the above results have been confirmed by numerical simulations, in particular through Molecular Dynamics (MD) and Direct simulation Monte Carlo (DSMC) [Bir94] methods. Those two numerical methods, although very different, show a surprisingly good agreement. This points out to correctness of the molecular chaos assumption and thus to the relevance of the DSMC method, which is particularly well adapted to simulate the dynamics of a homogeneous dilute gas.

3 Total Energy Fluctuations in Vibrated and Driven Granular Gases

3.1 The Inhomogeneous Boundary Driven Gas

In this section we will study the energy fluctuations of a granular gas in the case where the energy is injected into the system by a vibrating wall. Recently Aumaître et al. [AFFM04] investigated, by means of Molecular Dynamic (MD), the fluctuations of the total energy of the system. In particular they looked at the behavior of the first two moments of the energy pdf when the system size is changed, at constant averaged density. Because of the inhomogeneities, the mean kinetic energy is no more proportional to the number of particles, and thus it is not an extensive quantity, and analogously the mean kinetic temperature is no more intensive. This has led to the definition of an effective (intensive) temperature and an effective number of particles, which makes the energy extensive. In the following we will show how a rough calculation (neglecting correlations and small non-Gaussianity) using the hydrodynamic prediction for the temperature profile [VPB+06b], can explain the phenomenology observed in [AFFM04]. Within this description it is possible to get an expression of the effective temperature and number of particles as a function of the system parameters (i.e. number of particles, restitution coefficient, and temperature of the vibrating wall).

Energy Probability Distribution Function

In this part we will compute the energy pdf for a granular gas between two (infinite) parallel walls. The distance between the two walls is denoted by H, oriented along the x axis. Here we assume that one of the walls (in $x = 0$) has small and random vibrations, acting as a thermostat that fixes to T_0 the temperature at $x = 0$. Our boundary conditions therefore are:

$$T(\ell = 0) = T_0, \qquad \left.\frac{\partial T}{\partial \ell}\right|_{\ell=\ell_m} = \left.\frac{\partial T}{\partial x}\right|_{x=H} = 0, \tag{21}$$

where the rescaled length ℓ will be defined below. For the particular case of a steady state without macroscopic velocity flow, is it possible to solve those balance equations and get the temperature profile [BRMM00]:

$$T(\ell) = T_0 \left(\frac{\cosh\left(\sqrt{a(\alpha)}(\ell_m - \ell)\right)}{\cosh\left(\sqrt{a(\alpha)}\ell_m\right)} \right)^2, \tag{22}$$

where $a(\alpha)$ is a function of the restitution coefficient (its complete expression is given in [BdSMRM04]). The variable ℓ is proportional to the integrated density of the system on the x axis. Its definition is given by the following relation involving the local mean-free-path $\lambda(x)$:

$$d\ell = \frac{dx}{\lambda(x)}, \lambda(x) = \left[\frac{\sqrt{2}\pi^{\frac{d-1}{2}}}{\Gamma[(d+1)/2]} \sigma^{d-1} n(x) \right]^{-1}. \tag{23}$$

In the following we will suppose the velocity distribution to be a Maxwellian (a small non-Gaussian behavior exists, but it is not relevant for this calculation) with a local temperature (variance) given by (22):

$$f(\mathbf{v}, \ell) = \frac{e^{-\frac{v^2}{2T(\ell)}}}{(2\pi T(\ell))^{d/2}}. \tag{24}$$

The distribution for the energy of one particle ($e = v^2/2$) is hence:

$$p(e, T(\ell)) = f_{\frac{1}{T(\ell)}, \frac{d}{2}}(e), \tag{25}$$

where $f_{\alpha,\nu}(x)$ is the gamma distribution [Fel71]:

$$f_{\alpha,\nu}(x) = \frac{\alpha^\nu}{\Gamma(\nu)} x^{\nu-1} e^{-\alpha x}. \tag{26}$$

Our interest goes to the macroscopic fluctuations integrated over all the system. Thus, the macroscopic variable of interest is the granular temperature T_g, defined here as the average of the local temperature over the x profile:

$$T_g = \frac{1}{N} \int_V n(\mathbf{r}) T(\mathbf{r}) \, d\mathbf{r} = \frac{1}{\ell_m} \int_0^{\ell_m} T(\ell) \, d\ell. \tag{27}$$

with

$$\ell_m = N_x \frac{\sqrt{2}\pi^{\frac{d-1}{2}} \sigma^{d-1}}{\Gamma[(d+1)/2]}, \quad N_x = \frac{N}{V_{d-1}}, \tag{28}$$

where V_{d-1} is the area of the surface of dimension $d-1$ orthogonal to the x-direction, i.e. $H \times V_{d-1} = V$. When $d = 2$ one has $V_{d-1} \equiv L$ where L is the width of the system. N_x is the number of particles per unit of section perpendicular to the x axis. To get an expression of the energy pdf over the whole system, it is useful to divide the box in $\ell_m/\Delta\ell$ boxes of equal height (in the ℓ scale) $\Delta\ell$. It is helpful to use the length scale ℓ because the number of particles N_ℓ in each box of size $L \times \Delta\ell$ is a constant. Moreover, in each box i we will suppose the temperature a constant $T_i \equiv T(i\,\Delta\ell)$, defined expanding the granular temperature in a Riemann sum:

$$T_g = \lim_{\Delta\ell \to 0} \sum_{i=0}^{\ell_m/\Delta\ell} T_i\,\Delta\ell\,. \tag{29}$$

The calculation of the pdf of the box energy ϵ_i, i.e. a sum of the energies of the N_ℓ particles in a box i, is hence straightforward when the velocities of the particles are supposed to be uncorrelated:

$$q_i(y) \equiv \mathrm{prob}(\epsilon_i = y) = f_{\frac{1}{T_i},\frac{dN_\ell}{2}}(y), \tag{30}$$

The characteristic function of $q_i(y)$ is

$$\tilde{q}_i(k) = \frac{1}{(1 - ikT_i)^{\frac{dN_\ell}{2}}}\,. \tag{31}$$

Thus, the characteristic function for the kinetic energy of the whole system $E = \sum \epsilon_i$ can be obtained as the product of the characteristic functions $\tilde{q}_i(k)$:

$$\widetilde{P}(k) = \prod_{j=0}^{\ell_m/\Delta\ell} \tilde{q}_j(k) = \prod_{j=0}^{\ell_m/\Delta\ell} \frac{1}{(1 - ikT_j)^{\frac{dN_\ell}{2}}}. \tag{32}$$

Since the number of particle per box N_ℓ is a known fraction of the total number of particles ($N_\ell = N\Delta\ell/\ell_m$), one can rewrite the expression (32) as a Riemann sum. In the limit $\Delta\ell \to 0$ this yields the total kinetic energy characteristic function:

$$\widetilde{P}(k) = \exp\left(-\frac{dN}{2\ell_m} \int_0^{\ell_m} \log\left(1 - ikT(\ell)\right)\,d\ell\right). \tag{33}$$

Note that this result is valid for any temperature profile $T(\ell)$ and hence it can be applied also to other situations with different boundary conditions or different hydrodynamic equations.

Comparison with Simulations

Aumaître et al. [AFMP01, AFFM04] showed by Molecular dynamic simulations, that the pdf of the total energy is well fitted by a χ^2 law $\Pi(E) = f_{\frac{1}{T_E},\frac{N_f}{2}}(E)$ with a number of degrees of freedom N_f different from dN, and a temperature T_E different from the granular temperature T_g. The two parameters N_f and T_E are functions of the first two cumulants of the pdf:

$$N_f = 2 \frac{\langle E \rangle_c^2}{\langle E^2 \rangle_c}, \quad T_E = \frac{\langle E^2 \rangle_c}{\langle E \rangle_c}. \tag{34}$$

The notation $\langle X \rangle_c$ denotes the cumulant of the variable X. Here we want to compare result (33) with these numerical results. Since we are not able to analytically calculate the inverse Fourier Transform of (33) using (22) as a temperature profile, we used a numerical computation to obtain it in an approximate form. Moreover, an expression of the cumulants of the total kinetic energy can be obtained from the characteristic function (33):

$$\langle E^p \rangle_c = \frac{dN}{2\ell_m} \int_0^{\ell_m} T^p(\ell) d\ell. \tag{35}$$

In Fig. 1 the Inverse Fourier Transform of (33) is compared with the function $\Pi(E)$ previously defined. The similarity of the two functions is remarkable. Another important feature that can be checked with this results is the dependence of the above defined two macroscopic quantities (N_f and T_E) with system size. It is straightforward to see, from (27) and (35), that the granular temperature and the total kinetic energy are respectively an intensive and an extensive variable if ℓ_m is independent from the system size. This is effectively the case if both the density $\rho = N/V$ and the total height H are kept constant. Moreover, for large enough ℓ_m, the integral in (35) becomes size independent:

$$\int_0^{\ell_m} T^p(\ell) d\ell \sim \frac{T_0^p}{2p\sqrt{a(\alpha)}}. \tag{36}$$

Thus, the effective temperature T_E defined above becomes a constant proportional to the temperature of the wall, while the parameter N_f still depends on the system size:

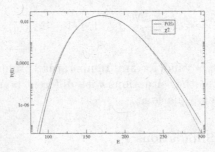

Fig. 1. Energy pdf (*solid line*) and a gamma distribution with same mean and same variance (*dotted line*) for a restitution coefficient $\alpha = 0.9$, $N = 100$ particles in two dimensions in a box of density $\rho = 0.04$, height $H = 50$, and with a wall temperature $T_0 = 5$

$$N_f \sim \frac{1}{\sqrt{a(\alpha)}} \frac{dN}{\ell_m}, \quad T_E \sim \frac{T_0}{2}. \tag{37}$$

Numerical simulations show that T_E effectively remains a constant for large systems, and under several procedures of box size increase. The behavior of N_f is determined by the maximum of the integrated density ℓ_m. For a square cell at constant density one finds $\ell_m \propto \sqrt{N}$, so that $N_f \propto \sqrt{N}$, which is not far from $N^{0.4}$ observed in [AFFM04]. Moreover, if only the height H of the cell is increased, ℓ_m is proportional to N, and N_f becomes constant. All those features are in agreement with the numerical observations in [AFFM04]. The above results clearly show that a rough calculation, which takes into account only the inhomogeneities of the system, is able to quantitatively describe the behavior of the fluctuations of the total kinetic energy of a vibrated granular gas. In some cases the energy pdf can be approximated with a gamma distribution, which is the standard distribution for the energy pdf in the canonical equilibrium. Nevertheless there are strong deviations from the equilibrium theory of fluctuations, since the two parameters of the gamma distribution (i.e. the temperature and the number of degrees of freedom) are not the granular temperature neither the number of degrees of freedom. Another important remark is that correlations, and in particular contributions from the two points distribution function, do not play a primary role to explain those deviations from the equilibrium theory of fluctuations. In order to characterize corrections arising from the two particles velocity pdf, one should measure energy fluctuations at a given height x from the vibrating wall. As already noted in [AFFM04] this task is very hard, since the available statistic become very poor. Nevertheless an effective way to quantify those fluctuations is to look at homogeneous systems, where contributions coming from the inhomogeneities vanish. With this objective in mind, we will be interested in the following in granular gases heated by an homogeneous and isotropic driving.

3.2 The Homogeneously Driven Case

In this section we will present some numerical results concerning the energy fluctuations in a dilute gas driven by the stochastic thermostat presented in Sect. 2.2 [VPB$^+$06b]. When the system reaches a stationary state, the dissipated energy is compensated by the energy injected by the thermostat, and the temperature fluctuates around its mean value. Here we are interested in the fluctuations of the total energy measured by the quantity

$$\sigma_E^2 = N \frac{\langle E^2(t) \rangle - \langle E(t) \rangle^2}{\langle E(t) \rangle^2}. \tag{38}$$

Note that $\sigma_E^2 \equiv 2N/N_f$. Brey et al. have computed, by means of kinetic equations, an analytical expression for σ_E^2 in the homogeneous cooling state, which is equivalent to the so-called Gaussian deterministic thermostat. One

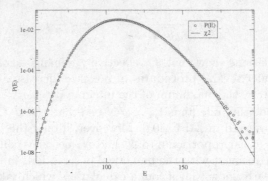

Fig. 2. Energy pdf (*dots*) from DSMC simulations with a restitution coefficient $\alpha = 0.5$ and $N = 100$ particles for a system driven with the stochastic thermostat. The *solid line* shows a gamma distribution with same mean and same variance

of the main differences of this stochastic thermostat with a deterministic one, is found in the elastic limit. On the one hand, for the cooling state, when the restitution coefficient tends to 1, the conservation of energy imposes that the energy pdf is a Delta function, and the quantity σ_E goes to 0. On the other hand, with the stochastic thermostat, if the elastic limit is taken keeping the temperature constant, the strength of the white noise will tend to zero, but it will still play a role in the velocity correlation function.

We performed DSMC simulations to measure the energy pdf of such a system. A plot of this pdf is shown in Fig. 2, and it is close to a χ^2-distribution with same mean and same variance. Nevertheless the number of degrees of freedom of this χ^2-distribution is lower than the true number of degrees of freedom (i.e. $(N-1) \times d$). This effect may arise from two separated causes: the non-gaussianity of the velocity pdf, and the presence of correlations between the velocities. This feature also suggests that a calculation of the energy pdf with the hypothesis of uncorrelated velocities (but non-Gaussian) could explain at least a part of this non-trivial effect. In order to quantify these contributions we will consider that the velocity pdf is well described by a Gaussian multiplied by the second Sonine polynomial:

$$f(v) = \frac{e^{-\frac{v^2}{2T}}}{(2\pi T)^{d/2}} \left(1 + a_2 S_2 \left(\frac{v^2}{2T}\right)\right), \tag{39}$$

where a_2 is given by expression (20).

The calculation of the pdf of the sum of the square of N variables distributed following (39) is straightforward. The characteristic function of the energy pdf is:

$$\widetilde{P}_N(k) = \frac{1}{(1 - ikT)^{\frac{Nd}{2}}} \left(1 + \frac{d(d+2)}{8} a_2 \left(\frac{1}{(1 - ikT)^2} - \frac{2}{(1 - ikT)} + 1\right)\right)^N \tag{40}$$

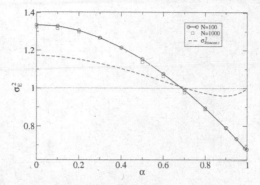

Fig. 3. Plot of σ_E^2 versus the restitution coefficient α for $N = 100$ (*open circle*) and $N = 1,000$ (*open square*) particles. The result of the calculation assuming uncorrelated velocities (42) is shown by the *dashed line*

where N is the number of particles of the system. This yields:

$$\langle E \rangle = \frac{d}{2} N T, \qquad \langle E^2 \rangle - \langle E \rangle^2 = \frac{d}{2} N T^2 \left(1 + \frac{d+2}{2} a_2 \right). \qquad (41)$$

It is now possible to have an explicit expression for the energy fluctuations:

$$\sigma_{E_{(uncorr.)}}^2 = \frac{2}{d} \left(1 + \frac{d+2}{2} a_2 \right). \qquad (42)$$

In Fig. 3 this result is compared with the result of DSMC simulations, performed for several values of the restitution coefficient α and for two different values for the number of particles N. The disagreement between the uncorrelated calculation and the simulations is a clear sign of the correlations induced by the inelasticity of the system. One can note that the fluctuations increase when the restitution coefficient decreases. One can also see that there is a value of the restitution coefficient α around $1/\sqrt{2}$, that is when the approximate expression of a_2 vanishes, for which σ_E^2 is exactly $1 \equiv 2/d$, as for a gas in the canonical equilibrium (velocities are then uncorrelated).

We now turn to the dependence of σ_E^2 on the strength of the white noise Γ. It is useful, for this purpose, to introduce a rescaled, dimensionless energy

$$\widetilde{E} = \frac{E - \langle E \rangle}{\sqrt{\langle E^2 \rangle - \langle E \rangle^2}}. \qquad (43)$$

We have plotted in Fig. 4 this rescaled energy pdf for a system of $N = 100$ particles with a restitution coefficient $\alpha = 0.5$ and for several values of the strength of the white noise Γ. One can see how all the pdfs collapse into a unique distribution. The role of the noise's strength is thus only to set the temperature (or mean kinetic energy) scale. Besides, relative energy fluctuations depend only on α and N. Moreover, since σ_E^2 does not depend on the

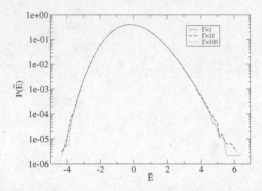

Fig. 4. Plot of the pdf of the rescaled energy $\widetilde{E}$ for a restitution coefficient $\alpha = 0.5$ and for $N = 100$ for several values of the strength of the noise Γ

number of particles N (for N large enough), the central limit theorem applies, and hence $P(\widetilde{E})$ is a Gaussian in the thermodynamic ($N \to \infty$) limit. In conclusion we have shown that randomly driven granular gases display non trivial fluctuations, because of the correlations induced by the inelasticity. Two different kinds of correlations contribute to this behavior of the fluctuations. First, the non-Gaussianity of the velocity pdf, which simply tells that the Euclidean components of the velocity of each particle are correlated one to each other. Second, a contribution from the two particles velocity pdf, which does not factorize exactly as a product of two one-particle distributions. It must be pointed out, however, that these correlations do not invalidate the Boltzmann equation. As already noted in [EC81, BdSMRM04, CPM07], the two points correlation function $g_2(\mathbf{v}_1, \mathbf{v}_2)$, which is defined by:

$$g_2(\mathbf{v}_1, \mathbf{v}_2) = f^{(2)}(\mathbf{v}_1, \mathbf{v}_2) - f(\mathbf{v}_1)f(\mathbf{v}_2), \tag{44}$$

where $f^{(2)}$ is the two points distribution, is of higher order in the density expansion (roughly speaking $\mathcal{O}(g(\mathbf{v}_1, \mathbf{v}_2)) \sim \mathcal{O}(f(\mathbf{v}_1)f(\mathbf{v}_2))/N$). This is confirmed by the numerical observations, since when the number of particles increases, the energy pdf tends be closer and closer to a Gaussian. Spatial correlations, which can be at work in homogeneously driven granular gases, at higher densities, and which have been neglected here (assuming spatial homogeneity) can also play a relevant role in fluctuations [PBV07].

4 A Large Deviation Theory for the Injected Power Fluctuations in the Homogeneous Driven Granular Gas

From now on we turn our attention to the fluctuations of another global quantity, i.e. the power injected into the system by the external source of energy. In particular in this section our main goal is to obtain a kinetic equation able to describe the behavior of the large deviations function of the time integrated injected power [VPB+05, VPB+06a] in a randomly driven gas (cf. Sect. 2.2). The latter quantity is the total work $\mathcal{W}$ provided by the thermostat over a time interval $[0, t]$:

$$\mathcal{W}(t) = \int_0^t dt \sum_i \mathbf{F}_i^{th} \cdot \mathbf{v}_i. \tag{45}$$

Our interest goes to the distribution of $\mathcal{W}(t)$, denoted by $P(\mathcal{W}, t)$, and to its associated large deviation function $\pi_\infty(w)$ defined for the reduced variable $w = \mathcal{W}/t$ ($\mathcal{W}(t)$ being extensive in time):

$$\pi_\infty(w) = \lim_{t \to \infty} \pi_t(w), \pi_t(w) = \frac{1}{t} \log P(\mathcal{W} = wt, t). \tag{46}$$

We introduce $\rho(\Gamma_N, \mathcal{W}, t)$ the probability that the system is in state Γ_N at time t with $\mathcal{W}(t) = \mathcal{W}$. The function we want to calculate is

$$P(\mathcal{W}, t) = \int d\Gamma_N \rho(\Gamma_N, \mathcal{W}, t). \tag{47}$$

We shall focus on the generating function of the phase space density

$$\hat{\rho}(\Gamma_N, \lambda, t) = \int d\mathcal{W} e^{-\lambda \mathcal{W}} \rho(\Gamma_N, \mathcal{W}, t) \tag{48}$$

and on the large deviation function of

$$\hat{P}(\lambda, t) = \int d\mathcal{W} e^{-\lambda \mathcal{W}} P(\mathcal{W}, t) = \int d\Gamma_N \hat{\rho}(\Gamma_N, \lambda, t) \tag{49}$$

which we define as

$$\mu(\lambda) = \lim_{t \to \infty} \frac{1}{t} \log \hat{P}(\lambda, t). \tag{50}$$

Note that $\mu(\lambda)$ is the generating function of the cumulants of $\mathcal{W}$, namely

$$\lim_{t \to \infty} \frac{\langle \mathcal{W}^n \rangle_c}{t} = (-1)^n \left. \frac{d^n \mu(\lambda)}{d\lambda^n} \right|_{\lambda=0}. \tag{51}$$

Moreover $\pi_\infty(w)$ can be obtained from $\mu(\lambda)$ by means of a Legendre transform, i.e. $\pi_\infty(w) = \mu(\lambda_*) + \lambda_* w$ with λ_* such that $\mu'(\lambda_*) = -w$.

The observable $\mathcal{W}$ is non-stationary but it is Markovian, hence a generalized Liouville equation for the extended phase-space density $\rho(\Gamma_N, \mathcal{W}, t)$ can be written. It varies in time under the combined effect of the inelastic collisions (which do not alter $\mathcal{W}$) and of the random kicks:

$$\partial_t \rho = \partial_t \rho \big|_{\text{collisions}} + \partial_t \rho \big|_{\text{kicks}} \tag{52}$$

Considering that the thermostat acts independently on each particle, it can be shown that

$$\partial_t \hat\rho\Big|_{\text{kicks}} = \sum_i \Big[\Gamma(\Delta_{\mathbf{v}_i} + 2\lambda\Gamma\mathbf{v}_i \cdot \partial_{\mathbf{v}_i} + \Gamma(d\lambda + \lambda^2 v_i^2)\Big]\hat\rho \qquad (53)$$

This additional piece is linear in $\hat\rho$ just as the collision part is. The large time behavior of $\hat\rho$ is governed by the largest eigenvalue $\mu(\lambda)$ of the evolution operator of $\hat\rho$. In the large time limit, we thus expect that

$$\hat\rho(\Gamma_N, \lambda, t) \simeq C(\lambda)e^{\mu(\lambda)t}\tilde\rho(\Gamma_N, \lambda), \qquad (54)$$

where $\tilde\rho(\Gamma_N, \lambda)$ is the eigenfunction associated to μ, and $C(\lambda)$ is such that $\tilde\rho(\Gamma_N, \lambda)$ is normalized to unity. We then introduce

$$\hat{f}^{(k)}(v_1, \ldots, v_k, \lambda, t) = \int \mathrm{d}\Gamma_{N-k}\hat\rho, \qquad (55)$$

where $\int \mathrm{d}\Gamma_{N-k}$ means an integration over $N - k$ particles, we have that

$$\partial_t \hat{f}^{(1)}(v, \lambda, t) = \Gamma\Delta_{\mathbf{v}}\hat{f} + 2\lambda\Gamma\partial_{\mathbf{v}} \cdot \mathbf{v}\hat{f} + \Gamma(\lambda^2 v^2 - d\lambda)\hat{f} + \hat{J} \qquad (56)$$

with $\hat{J} = \int \mathrm{d}\mathcal{W}e^{-\lambda\mathcal{W}}J$ the Laplace transform of the collision integral in which $f(v, \mathcal{W}, t)$ now plays the rôle of the velocity distribution. Quite unexpectedly the above equation has a straight physical interpretation: consider a many particle system where a noise of strength Γ and a viscous friction-like force $\mathbf{F} = -2\lambda\Gamma\mathbf{v}$ act independently on each particle, and where the particles interact by inelastic collisions. Consider then that the particles annihilate/branch (depending on the sign of λ) at constant rate $d\lambda\Gamma$, and branch with a rate proportional to $\lambda^2 v^2\Gamma$. Then, the equation governing the evolution of the one particle velocity distribution of such a system is exactly (56), where λ is a parameter tuning the strength of the external fields. Moreover, in spite of there being no a priori reason for that, $\tilde\rho$, as well as $\tilde{f} = \int \mathrm{d}\Gamma_{N-1}\tilde\rho$, can be interpreted as probability density functions.

The one and two-point functions $f^{(1)}(v, \mathcal{W}, t)$ and $f^{(2)}(v_1, v_2, \mathcal{W}, t)$ that enter the expression of J are expected to verify, at large times,

$$\hat{f}^{(1)}(v_1, \lambda, t) = C(\lambda)e^{\mu t}\tilde{f}^{(1)}(v_1, \lambda), \qquad (57)$$

and

$$\hat{f}^{(2)}(v_1, v_2, \lambda, t) = C(\lambda)e^{\mu t}\tilde{f}^{(2)}(v_1, v_2, \lambda), \qquad (58)$$

where both $\tilde{f}^{(1)}$ and $\tilde{f}^{(2)}$ are normalized to unity. We perform the following molecular-chaos-like assumption:

$$\tilde{f}^{(2)}(v_1, v_2, \lambda) \simeq \tilde{f}^{(1)}(v_1, \lambda)\tilde{f}^{(1)}(v_2, \lambda) \qquad (59)$$

which does have a definite physical interpretation in the language of the inelastic hard-spheres with fictitious dynamics (viscous friction, velocity dependent branching/annihilation) described in the above paragraph. Then we get that

$$\mu \tilde{f}(v, \lambda) = \Gamma \Delta_{\mathbf{v}} \tilde{f} + 2\lambda \Gamma \mathbf{v} \cdot \partial_{\mathbf{v}} \tilde{f} + \Gamma (d\lambda + \lambda^2 v^2) \tilde{f}$$
$$+ \frac{1}{\ell} \int_{\mathbf{v}_{12} \cdot \hat{\sigma} > 0} d v_2 d \hat{\sigma} \mathbf{v}_{12} \cdot \hat{\sigma} \left[\alpha^{-2} \tilde{f}(v_1^{**}, \lambda) \tilde{f}(v_2^{**}, \lambda) - \tilde{f}(v_1, \lambda) \tilde{f}(v_2, \lambda) \right]$$

$$(60)$$

where we have now omitted the superscript (1) denoting the one-point function. The $\lambda = 0$ limiting case yields the usual Boltzmann equation, since in this case a stationary solution exists, and hence $\mu(\lambda = 0) = 0$. The boundary condition to the evolution equation above is thus:

$$\tilde{f}(\mathbf{v}, \lambda = 0) = f_{st}(\mathbf{v}) \tag{61}$$

with $f_{st}(\mathbf{v})$ the stationary velocity pdf (cf. (17)).

4.1 The Cumulants

Here we find an approximated expression of $\mu(\lambda)$ solving a system of equations obtained projecting (60) on the first velocity moments. First we shall define a dimensionless velocity $\mathbf{c} = \mathbf{v}/v_0(\lambda)$, where $v_0(\lambda)$ plays the role of a thermal velocity:

$$v_0^2(\lambda) = 2T(\lambda) = \frac{2}{d} \int d\mathbf{v} \, v^2 \, \tilde{f}(v, \lambda). \tag{62}$$

Then, defining the function $f(\mathbf{c}, \lambda) = v_0(\lambda) \tilde{f}(\mathbf{v}, \lambda)$, and its related moments of order n

$$m_n(\lambda) = \int d\mathbf{c} \, c^n f(c, \lambda), \tag{63}$$

one obtains the following recursion relation:

$$(\mu + \Gamma(2n + d)\lambda) m_n = \frac{\Gamma}{v_0^2} n(n + d - 2) m_{n-2} + \Gamma \lambda^2 v_0^2 m_{n+2} - v_0 \nu_n, \quad (64)$$

where

$$\nu_n = - \int d\mathbf{c} \, c^n \, J[f, f]. \tag{65}$$

Recalling the definition of the cumulants (51), and the approximated solution for the stationary velocity pdf, it appears natural to argue that, for $\lambda \sim 0$, the function $f(\mathbf{c}, \lambda)$ should be well approximated by:

$$f(c, \lambda) = \phi(c) \left(1 + a_1(\lambda) S_1 \left(c^2\right) + a_2(\lambda) S_2 \left(c^2\right)\right) + \mathcal{O}(a_3), \tag{66}$$

where $\phi(c) = \pi^{-d/2} \exp(-c^2)$ is the Gaussian distribution. Even in this case, from the relation (19) and from the definition (62), the coefficient a_1 is found to be 0. The method consists in taking (64) for $n = 0, 2$ and 4 in order to find an explicit expression of μ, v_0, and a_2 in the limit $\lambda \to 0$. The quantities ν_2

and ν_4 have been calculated at the first order in a_2 [vNE98], and their explicit expressions are:

$$\nu_2 = \frac{(1-\alpha^2)}{2\ell} \frac{\Omega_d}{\sqrt{2\pi}} \left\{ 1 + \frac{3}{16} a_2 \right\} = \frac{d\Gamma}{\sqrt{2T_0^3}} \left\{ 1 + \frac{3}{16} a_2 \right\}, \tag{67}$$

and

$$\nu_4 = \frac{d\Gamma}{\sqrt{2T_0^3}} \left\{ T_1 + a_2 T_2 \right\}, \tag{68}$$

with

$$T_1 = d + \frac{3}{2} + \alpha^2 \tag{69}$$

$$T_2 = \frac{3}{32}(10\,d + 39 + 10\,\alpha^2) + \frac{(d-1)}{(1-\alpha)}, \tag{70}$$

where $T_0 = \left(\frac{2d\Gamma\ell\sqrt{\pi}}{(1-\alpha^2)\Omega_d} \right)^{2/3}$ is the granular temperature obtained averaging over Gaussian velocity pdfs (i.e. the zero-th order of Sonine expansion). The expression of the first moments m_n is:

$$m_0 = 1 \tag{71a}$$

$$m_2 = d/2 \tag{71b}$$

$$m_4 = \frac{(1+a_2)\,d\,(2+d)}{4} \tag{71c}$$

$$m_6 = \frac{(1+3\,a_2)\,d\,(2+d)\,(4+d)}{8} \tag{71d}$$

With the help of the above defined temperature scale T_0, we introduce some dimensionless variables:

$$\tilde{\mu} = \mu \frac{T_0}{d\Gamma}, \qquad \tilde{\lambda} = \lambda T_0,$$

$$\tilde{v}_0^2 = \frac{v_0^2}{2T_0}, \qquad \tilde{\nu}_p = \frac{\sqrt{2T_0^3}}{\Gamma} \nu_p. \tag{72}$$

Note that this scaling naturally defines the scales for the other quantities of interest, namely:

$$\tilde{\pi}_t = \pi_t \frac{T_0}{d\Gamma}, \qquad \tilde{w} = \frac{w}{d\Gamma}, \qquad \tilde{\mathcal{W}} = \frac{\mathcal{W}}{\langle \mathcal{W} \rangle}. \tag{73}$$

The expression of the moment equation (64) becomes, for the above defined dimensionless quantities:

$$\left(\tilde{\mu} d + (2n+d)\tilde{\lambda} \right) m_n = \frac{n(n+d-2)}{2\tilde{v}_0^2} m_{n-2} + 2\tilde{v}_0^2 m_{n+2} - \tilde{v}_0 \tilde{\nu}_n. \tag{74}$$

First we solve the above equation for $n = 0$, getting the following result:

$$\tilde{\mu}(\tilde{\lambda}) = -\tilde{\lambda} + \tilde{\lambda}^2 \tilde{v}_0^2(\tilde{\lambda}).$$

(75)

Recalling that when $\lambda \to 0$ one has $v_0^2 = 2T_g + \mathcal{O}(\lambda)$, it is important to note that if we restrict our analysis to the Gaussian approximation for $P(\mathcal{W}, t)$, that is if we truncate $\mu(\lambda)$ to order λ^2, (75) will read:

$$\frac{\mu}{d\Gamma} = \lambda(\lambda T_g - 1).$$

(76)

Then we see that indeed

$$\mu(\lambda) = \mu\left(\frac{1}{T_g} - \lambda\right),$$

(77)

which means that $\pi_\infty(w) = \max_\lambda \{\mu(\lambda) + \lambda w\}$ verifies

$$\pi_\infty(w) - \pi_\infty(-w) = \frac{w}{T_g}.$$

(78)

However, the nontrivial functions $m_n(\lambda)$ will break the property (77), as we shall explicitly show later. In order to characterize more precisely the dependence of $\tilde{\mu}$ upon $\tilde{\lambda}$ for small values of $\tilde{\lambda}$, it is useful to expand $\tilde{v}_0^2$ and a_2 in powers of $\tilde{\lambda}$:

$$\tilde{v}_0^2(\tilde{\lambda}) = \tilde{v}_0^{2(0)} + \tilde{\lambda}\tilde{v}_0^{2(1)} + \tilde{\lambda}^2 \tilde{v}_0^{2(2)} + \mathcal{O}(\tilde{\lambda}^3)$$

(79a)

$$a_2(\tilde{\lambda}) = a_2^{(0)} + \tilde{\lambda}a_2^{(1)} + \tilde{\lambda}^2 u_2^{(2)} + \mathcal{O}(\tilde{\lambda}^3)$$

(79b)

In this way we can find $\tilde{v}_0^{2(i)}\left(a_2^{(i)}\right)$ solving (74) for $n = 2$:

$$\tilde{v}_0^{2(0)} = \left(1 - \frac{a_2^{(0)}}{8}\right),$$

(80)

$$\tilde{v}_0^{2(1)} = -\frac{4}{3} + \frac{a_2^{(0)}}{3} - \frac{a_2^{(1)}}{8},$$

(81)

$$\tilde{v}_0^{2(2)} = 2 - a_2^{(0)}\left(\frac{1}{12} + \frac{d}{3}\right) + \frac{a_2^{(1)}}{3} - \frac{a_2^{(2)}}{8}.$$

(82)

Then we substitute $\tilde{v}_0^2(\tilde{\lambda})$ in the third equation and expand it in powers of $\tilde{\lambda}$ to find the expression of $a_2^{(i)}(\alpha)$. Note that one has also to expand in powers of a_2 and keep only the linear terms in order to be coherent with the $\tilde{\nu}_p$ calculations. We find the following expressions, which are plotted in Fig. 5:

$$a_2^{(0)} = \frac{4(1-\alpha)(1-2\alpha^2)}{19 + 14d - 3\alpha(9 + 2d) + 6(1-\alpha)\alpha^2}$$

(83)

$$a_2^{(1)} = -\frac{4(1-\alpha)^2(-1+2\alpha^2)(31 + 2\alpha^2 + 16d)}{(19 + 14d - 3\alpha(9 + 2d) + 6(1-\alpha)\alpha^2)^2}$$

(84)

$$a_2^{(2)} = \frac{A(\alpha)}{B(\alpha)}$$

(85)

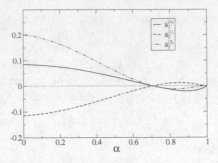

Fig. 5. $a_2^{(0)}$, $a_2^{(1)}$ and $a_2^{(2)}$ versus α for $d = 2$

with

$$
\begin{aligned}
A(\alpha) = {} & 16\left(-1+\alpha\right)^2\left(-1+2\,\alpha^2\right)\times \\
& \times\{906 + \alpha\left[-984 + \alpha\left(85 + 3\,\alpha\left(-19 + 6\left(-1+\alpha\right)\alpha\right)\right)\right] + 985\,d + \\
& + \alpha\left[-951 + \alpha\left(-25 + 3\,\alpha\left(7 + 6\left(-1+\alpha\right)\alpha\right)\right)\right]d + \\
& + \left(269 + 3\,\alpha\left(-75 + 2\,\alpha\left(-7 + 3\,\alpha\right)\right)\right)d^2\},
\end{aligned}
\tag{86}
$$

and

$$
B(\alpha) = 3\left(-19 - 14\,d + 3\,\alpha\left(9 + 2\left(-1+\alpha\right)\alpha + 2\,d\right)\right)^3
\tag{87}
$$

The $v_0^{2^{(0)}}$ expression, as well as the $a_2^{(0)}$ expression, gives the usual results established for granular gases [vNE98, MS00]. At this point the computation of the cumulants becomes straightforward. From relation (51) it follows:

$$
\lim_{t\to\infty} \frac{\langle \mathcal{W}^n\rangle_c}{t} = (-1)^n N d \Gamma T_0^{n-1} n! \, \tilde{v}_0^{2^{(n-2)}}.
\tag{88}
$$

Moreover, since the $a_2^{(i)}$ corrections are numerically small, the zero-th order (Gaussian) approximation already gives a good estimate for the cumulants. Namely, the first cumulants are , in this approximation:

$$
\begin{aligned}
\langle\mathcal{W}\rangle_c &= t N d\Gamma, & \langle\mathcal{W}^2\rangle_c &= 2t N d\Gamma T_0, \\
\langle\mathcal{W}^3\rangle_c &= 8t N d\Gamma T_0^2, & \langle\mathcal{W}^4\rangle_c &= 48t N d\Gamma T_0^3.
\end{aligned}
\tag{89}
$$

All the above expansions in powers of λ, at the second order in Sonine coefficients (e.g. a_2) can be carried out just expanding v_0 and a_2 in (79) to higher powers of λ. Moreover, expanding in higher order in Sonine coefficient (e.g. a_3) remains in principle still possible, but it will involve a higher number of equations in the hierarchy (74) (e.g. $n = 6$), and therefore will need the expression of higher order collisional moments (e.g. ν_6).

4.2 The Solvable Infinite Dimension Limit

Strong arguments [VPB$^+$06a] can be given showing that in high dimensions $\tilde{f}(v, \lambda)$ is not far from a Gaussian. We are therefore led to consider, in the limit $d \to \infty$, $\tilde{f}(\mathbf{v}, \lambda)$ to be a Gaussian with a λ-dependent second moment. In this situation the dimensionless function f will read:

$$f(\mathbf{c}) = \frac{e^{-c^2}}{\pi^{d/2}} \tag{90}$$

with $\mathbf{c} = \mathbf{v}/v_0(\lambda)$. In this context one can solve (74) in order to get an explicit expression for $\mu(\lambda)$. Solving the system defined by (74) for $n = 0$ and $n = 2$ gives a unique solution for $\tilde{\mu}(\tilde{\lambda})$ which verifies the physical requirement $\tilde{\mu}(0) = 0$:

$$\tilde{\mu}(\tilde{\lambda}) = -\tilde{\lambda} + \frac{\tilde{\lambda}^2}{2} \tilde{v}_0^2(\tilde{\lambda}), \tag{91}$$

with:

$$\tilde{v}_0^2(\tilde{\lambda}) = \frac{1 + 4\tilde{\lambda}^3}{2\tilde{\lambda}^4} + \frac{b_1(\tilde{\lambda})}{2} -$$

$$- \frac{1}{2} \left[-\frac{32}{\tilde{\lambda}^2} + \frac{2\left(1 + 4\tilde{\lambda}^3\right)^2}{\tilde{\lambda}^8} + b_2(\tilde{\lambda}) - b_3(\tilde{\lambda}) + \frac{b_4(\tilde{\lambda})}{4b_1(\tilde{\lambda})} \right]^{\frac{1}{2}}, \tag{92}$$

and

$$b_1(\tilde{\lambda}) = \sqrt{\tilde{\lambda}^{-8} + \frac{8}{\tilde{\lambda}^5} - b_2(\tilde{\lambda}) + b_3(\tilde{\lambda})},$$

$$b_2(\tilde{\lambda}) = \frac{16\left(\frac{2}{3}\right)^{\frac{1}{3}}}{\tilde{\lambda}^3 \left(9 + \sqrt{3}\sqrt{27 + 256\tilde{\lambda}^3}\right)^{\frac{1}{3}}},$$

$$b_3(\tilde{\lambda}) = \frac{2\left(\frac{2}{3}\right)^{\frac{2}{3}}\left(9 + \sqrt{3}\sqrt{27 + 256\tilde{\lambda}^3}\right)^{\frac{1}{3}}}{\tilde{\lambda}^4},$$

$$b_4(\tilde{\lambda}) = \frac{256}{\tilde{\lambda}^3} - \frac{192\left(1 + 4\tilde{\lambda}^3\right)}{\tilde{\lambda}^6} + \frac{8\left(1 + 4\tilde{\lambda}^3\right)^3}{\tilde{\lambda}^{12}}. \tag{93}$$

This expression of the velocity scale reduces to the kinetic temperature for $\lambda = 0$, and decreases monotonically as $\lambda^{-1/2}$ when $\lambda \to \infty$. This means that in the limit $\lambda \to \infty$ $\tilde{f}$ approaches a Dirac distribution as $\exp(-\lambda v^2/2)$. This feature supports the intuition that the small $\mathcal{W}$ events (which are related to the large values of λ) are provided by the small velocities. The behavior of $\tilde{\mu}$ is shown in Fig. 6. The large deviations function $\tilde{\mu}(\tilde{\lambda})$ becomes complex for $\tilde{\lambda} < -\frac{3}{2^{8/3}}$, because of the terms containing $\sqrt{27 + 256\tilde{\lambda}^3}$. Moreover for large

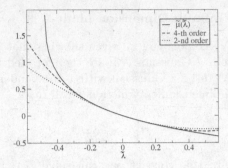

Fig. 6. The *solid line* shows $\tilde{\mu}$ in the limit $d \to \infty$. The *dashed line* is $\tilde{\mu}$ at fourth order in $\tilde{\lambda}$ from (75) for $d = 2$ and $\alpha = 0.5$. Finally the *dotted line* shows the same quantity calculated with a truncation at second order in λ, which would satisfy the G–C relation

$\tilde{\lambda}$ the behavior of this function is $\tilde{\mu}(\tilde{\lambda}) \sim -\tilde{\lambda}^{\frac{1}{4}}$. In the vicinity of the singularity (i.e. $\tilde{\lambda} = \lambda_0 = -\frac{3}{2^{8/3}}$) the behavior of the large deviation function is:

$$\tilde{\mu}(\tilde{\lambda}) = \frac{3}{2^{3/2}} - 3^{2/3} 2^{1/6} \sqrt{\tilde{\lambda} - \lambda_0} + \mathcal{O}(\tilde{\lambda} - \lambda_0). \tag{94}$$

From the behavior for large $\tilde{\lambda}$ it is possible to recover the left tail of the large deviation function π_∞. In general, if $\mu(\lambda) \sim -\lambda^\beta$ for $\lambda \to \infty$, this leads to $\mu'(\lambda_*) = -\beta\lambda_*^{\beta-1} = -w$. This last relation tells us that for $\beta < 1$ we are recovering the limit $w \to 0^+$, with a behavior of the large deviation function given by $\pi_\infty(w) = \mu(\lambda_*) + \lambda_* w \sim w^{\frac{\beta}{\beta-1}}$. Moreover, from the behavior of μ near λ_0, an analogous calculation provides the right tail of the large deviation function: $\pi_\infty(w) \sim \lambda_0 w$, when $w \to \infty$. Finally, in our particular case, the tails are given by

$$\tilde{\pi}_\infty(\tilde{w} \to 0^+) \sim -\tilde{w}^{-1/3}, \ \tilde{\pi}_\infty(\tilde{w} \to \infty) \sim -\tilde{w}, \tag{95}$$

Note that there is no $w < 0$ tail to $\tilde{\pi}_\infty$. The graph of the whole function $\tilde{\pi}_\infty(\tilde{w})$ is depicted in Fig. 7.

5 Fluctuations of Injected Power at Finite Times: Two Examples

5.1 The Homogeneous Driven Gas of Inelastic Hard Disks

In this section the results of numerical simulations of two models (inelastic hard spheres and inelastic Maxwell model) are presented with particular attention to the verification of the Fluctuation Relation for the injected power. The main requirement to pose the question about the validity of the Fluctuation Relation is a clean observation of a negative tail in the pdf of the injected

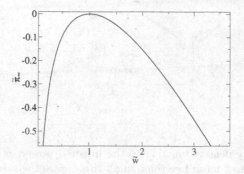

Fig. 7. $\tilde{\pi}_{\infty}(\tilde{w})$

power. This dramatically limits the time t of integration of $\mathcal{W}(t)$. In numerical simulations, as well as in real experiments, at time larger than a few mean free times the negative tail disappears. On the other hand, at times of the order of 1–3 mean free times, the Fluctuation Relation appears to be correctly verified for the inelastic Hard Spheres model and slightly violated for the inelastic Maxwell model. The measure of the cumulants, anyway, gives a neat indication of the fact that the time of convergence of the large deviation function is at least 10 times as large and that the true asymptotic is well reproduced by the theory exposed in this chapter. This theory shows strong arguments against the validity of a symmetry relation of the Gallavotti–Cohen type for the large deviations of injected power.

The stationary state of a driven granular gas, modeled by (16), under the assumption of Molecular Chaos may be studied with a Direct Simulation Monte Carlo technique [Bir94, MS00]. As a first check of reliability of the algorithm, we have measured the granular temperature T_g and the first non-zero Sonine coefficient $a_2 \equiv (\langle v^4 \rangle / \langle v^2 \rangle^2 - 3)/3$. The measured granular temperature is always in perfect agreement with the estimate. The measured a_2 coefficient is a highly fluctuating quantity and its average is in very good agreement with the theoretical estimate.

In Fig. 8 the probability density functions $p(w,t) \equiv tP(wt,t)$ (for t equal to 1 mean free time) for three different choices of parameters N, Γ (at fixed restitution coefficient α) is shown. The values of the first two cumulants of the distribution and their theoretical values are compared in Table 1, with very good agreement. In the same table we present also the measure of the third and fourth cumulants.

The comparison with a Gaussian with same mean value and same variance shows that the pdf $P(\mathcal{W}, t)$ is not exactly a Gaussian. In particular there are deviations from the Gaussian form in the right (positive) tail. This is well seen in Fig. 9. It must be noted that the important deviations in the right tail arise at values of $\mathcal{W}(t)$ larger than the minimum $\mathcal{W}(t)$ available in the left tail, i.e. they have no influence in the following plot of Fig. 10 regarding the Gallavotti–Cohen symmetry.

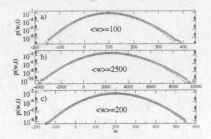

Fig. 8. Probability density function of the injected power, $p(w, t) \equiv tP(\mathcal{W}(t) = wt, t)$ with t equal to 1 mean free time. In all three cases the value of the restitution coefficient is $\alpha = 0.9$. Other parameters are (**a**) $N = 100$, $\Gamma = 0.5$; (**b**) N = 100, $\Gamma = 12.5$; (**c**) $N = 200$, $\Gamma = 0.5$. The *dashed line* represents a Gaussian with same first two cumulants. These distributions have been obtained with $\sim 1.5 \times 10^9$ independent values of $\mathcal{W}(t)$

Table 1. Rescaled cumulants of the distribution of injected work $P(\mathcal{W}, t)$, measured with t equal to 1 mean free time for different choices of the parameters

N	Γ	$\langle \mathcal{W}(t) \rangle / t$	$\langle \mathcal{W}(t)^2 \rangle_c / t$	$N\Gamma d$	$2N\Gamma d T_g$	$\langle \mathcal{W}(t)^3 \rangle_c / t$	$\langle \mathcal{W}(t)^4 \rangle_c / t$
100	0.5	100	20,835	100	21,052	6.02779×10^5	1.54181×10^8
100	12.5	2,500	13,019,125	2,500	13,157,900	9.47684×10^9	6.12963×10^{13}
200	0.5	199.9	42,009	200	42,120	1.21911×10^6	3.09634×10^8

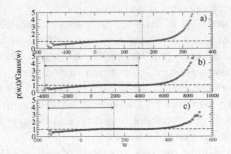

Fig. 9. Ratio of $P(\mathcal{W}, t)$ and a Gaussian with the same first two moments, for the same parameters as in Fig. 8: (**a**) corresponds to $N = 100$, $\Gamma = 0.5$, (**b**) to $N = 100$, $\Gamma = 12.5$ and (**c**) to $N = 200$, $\Gamma = 0.5$. The range between the *vertical dotted lines* is the useful one for the check of the Gallavotti–Cohen relation. It can be noted that the strongest deviations from the Gaussian behavior appear outside of this range

In Fig. 10 the Gallavotti–Cohen relation $\pi_t(w) - \pi_t(-w) = \beta_{eff} w$ is questioned for the same choice of the parameters. The relation, at this level of resolution and for this value of the time t (1 mean free time), is well satisfied. Moreover Table 2 shows that the value of β_{eff} is well approximated by $\beta = 1/T_g$, as expected if the truncation of $\mu(\lambda)$ at the second order were valid, see (78). In Fig. 11 the same relation is checked for different values of t,

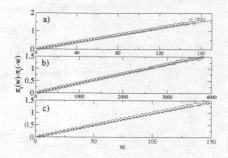

Fig. 10. Plot of $\pi_t(w) - \pi_t(-w)$ versus w, for the same choice of the parameters as in Fig. 8. The values of the slope β_{eff} of the fitting *dashed lines* are in Table 2

Table 2. Factor of proportionality in the "Gallavotti–Cohen" relation compared with β

N	Γ	β_{eff}	$1/T_g$
100	0.5	0.0100	0.00955
100	12.5	0.000402	0.000382
200	0.5	0.00995	0.00952

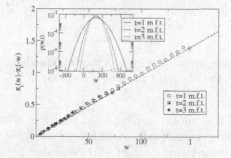

Fig. 11. Plot of $\pi_t(w) - \pi_t(-w)$ versus w, for the system with $N = 100$ and $\Gamma = 0.5$ for different values of t. We recall that in this case $\langle w \rangle = 100$. The *dashed line* has slope $\beta = 1/T_g$. In the inset the corresponding $p(w,t)$ are shown

slightly larger (i.e. up to t equal to 3 mean free times). No relevant deviations are observed as t is increased. Moreover this figure is important to understand the dramatic consequences that a larger t has on the "visibility" of the Gallavotti–Cohen symmetry: as t is increased, events with negative integrated power injection become rarer and rarer. This eventually leads to the vanishing of the left branch of $P(\mathcal{W}, t)$.

The main conclusion is that no appreciable departure from the λ^2 truncation is observed at this level of resolution. Much larger statistics are required to probe the very high energy tails of $p(w,t)$. Further numerical insights make evident that the small times used to check the GC Relation (t smaller or

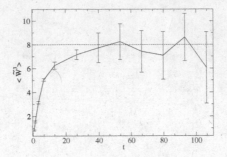

Fig. 12. Third cumulant $\tilde{\mathcal{W}} = \mathcal{W}/(\langle \mathcal{W} \rangle T_g^2)$ for several times of integration. The time is in units of the mean collision time. Note that the time when a stationary value of the rescaled cumulant is reached is much larger than the characteristic time of the system (the collision time)

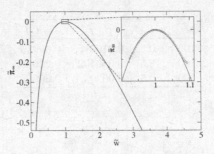

Fig. 13. Numerical measurement of $\tilde{\pi}_t$ for a time of 50 collisions per particle (when a stationary value for the rescaled cumulant is reached)

equal than 3 mean free times) are far from the time where the asymptotic large deviation scaling starts working. In Fig. 12 we show indeed the numerical measure of the third cumulant of $\mathcal{W}(t)$ rescaled by the first cumulant, varying the integration time t. The time of saturation is of the order of ~ 50 mean free times. The saturation value is in very good agreement with the value predicted by our theory, (89). Note that this value is not at all trivial, since the third cumulant for a Gaussian distribution is zero. At that time the measurable $\pi_t(w)$ is shown in Fig. 13, rescaled by $\langle w \rangle$. The accessible range of values from a numerical simulation is dramatically poor and we think it is already remarkable to have obtained a good measure of the third cumulant with such a resolution.

The reason for a verification at small times of the GC formula is the following: near $w = 0$ the pdf of w is almost a Gaussian. In the Gaussian case we immediately get $\pi_t(w) - \pi_t(-w) = \beta_{eff} w$ with $\beta_{eff} = 2\langle \mathcal{W}(t) \rangle / \langle \mathcal{W}(t)^2 \rangle_c$. The first two cumulants at small times are easily obtained considering an uncorrelated sequence of energy injection, obtaining $\langle \mathcal{W}(t) \rangle / t = N\Gamma d$ and $\langle \mathcal{W}(t)^2 \rangle_c / t = \langle (\sum_i \mathbf{F}_i^{th} \cdot \mathbf{v}_i)^2 \rangle_c = 2N\Gamma d T_g$. Then the value $\beta_{eff} = 1/T_g$ is unavoidable. In this case the GCFR observed is nothing else than the

Green–Kubo (or Einstein) relation, which is known to be valid for driven granular gases: $\langle \mathcal{W}(t)^2 \rangle / t = 2T_g \langle \mathcal{W}(t) \rangle / t$ [PBL02, PBV07]. Small deviations from a Gaussian appear, in first approximation, as small deviations from the slope $1/T_g$, but the straight line behavior is robust since the first non-linear term of $\pi_t(w) - \pi_t(-w)$ is not w^2 but w^3 [AFMP01].

Numerical simulations of the Inelastic Maxwell Model have been performed with a Direct Simulation Monte Carlo analogous to the one used in the Hard Spheres model. The Maxwell gas is a kinetic model due to Maxwell, who observed that a pair potential proportional to $r^{-2(d-1)}$, r being the distance between two interacting particles, gives rise to a great simplification of the collision integral [Max67]. In fact this kind of interaction makes the collision frequency velocity independent. It must be noted that when the inelasticity of the particles is considered, this model looses its straight physical interpretation, but it nevertheless keeps its own interest. The collision integral is analytically simpler than the hard particles model and preserves the essential physical ingredients in order to have qualitatively the same phenomenology. In the recent development of granular gases this kinetic model has been extensively investigated [BMP02, BNK02, BNK03, EB02, BCG00]. Thanks to the simplifications present in this model, we are able to improve the number of collected data by more than a factor of ten. The distributions of the injected power $p(w, t)$ are shown in Fig. 14 for some choices of the restitution coefficient α. The driving amplitude Γ has been changed in order to keep constant the stationary granular temperature T_g. In Fig. 15 we have displayed the deviations from the Gaussian of $P(\mathcal{W}, t)$. The non-Gaussianity of $P(\mathcal{W}, t)$ is highly pronounced, but again it is striking only in the positive branch of the pdf. We have tried, with success, a fit with a fourth order polynomial, which is consistent with the usual truncation of the Sonine expansion to the second Sonine polynomial.

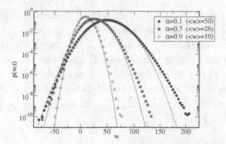

Fig. 14. $p(w, t) \equiv tP(wt, t)$ for different values of α (at fixed constant temperature T_g) in the Driven Inelastic Maxwell Model measured at a time t equal to 1 mean free time. The *dashed lines* are Gaussian distributions with the same mean and same variance. These distributions have been obtained with $\sim 4 \times 10^{10}$ independent values of $\mathcal{W}(t)$

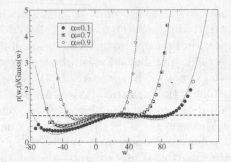

Fig. 15. $p(w,t)$ (at t equal to 1 mean free time) divided by a Gaussian with same average and same variance for different values of α (at fixed constant temperature T_g) in the Driven Inelastic Maxwell Model. The *light dashed lines* represent a fit with a polynomial of fourth order

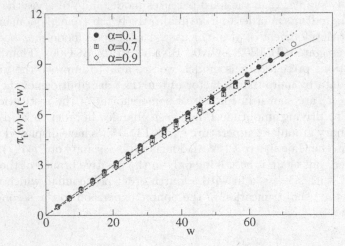

Fig. 16. (Color online). Finite time check of Gallavotti–Cohen relation for the injected power (with t equal to 1 mean free time), i.e. $\pi_t(w) - \pi_t(-w)$ versus w, in a numerical simulation of the Driven Inelastic Maxwell Model with $N = 50$, and different values of α (the driving amplitude Γ has been rescaled in order to fix the granular temperature T_g). The *dashed curve* is a straight line with slope $\beta = 1/T_g$. The *dotted curve* is a straight line obtained fitting the $\alpha = 0.1$ data points until $w = 45$, useful as a guide for the eye. The *thin solid curve* is a fit with a cubic $(0.28w + 5.6 \cdot 10^{-4}w^2 - 1.1 \cdot 10^{-5}w^3)$

Finally, in Fig. 16, we have attempted a check of the Gallavotti–Cohen fluctuation relation. The relation seems to be systematically violated. This appears in two points (1) the right–left ratio of the large deviation function is not a straight line; (2) the best fitting line has a slope which is larger than $1/T_g$. The "curvature" (and the deviation from the $1/T_g$ line) increases with decreasing values of α, indicating that the inelasticity is the cause of the deviation from the Gallavotti–Cohen relation. It should be noted that to achieve this result we have collected more than 4×10^{10} independent values of $\mathcal{W}(t)$, so that the statistics of the negative large deviations could be clearly displayed.

5.2 The Boundary Driven Gas of Inelastic Hard Disks

In a recent experiment on vibrated granular gases [FM04] it has been argued that the statistics of the power injected on a subsystem by the rest of the gas fulfills the Fluctuation Relation (FR) by Gallavotti and Cohen. The experiment was performed by putting in a two-dimensional vertical box N disks of glass and submitting the container to a strong vertical vibration. We have reproduced the experiment by means of a Molecular Dynamics (MD) simulations of inelastic hard disks, observing perfect agreement with the experimental results and obtaining a deeper insight into the system. The main difference of this model with respect to the previous "homogeneously driven" model is that the external energy source is located at the two horizontal (top and bottom) walls. This boundary driving mechanism leads to the development of spatial inhomogeneities and the appearance of internal currents.

The event driven MD simulations have been performed for a system of N inelastic hard disks with restitution coefficient α, diameter σ and mass $m = 1$. The vertical two-dimensional box of width $L_x = 48\sigma$ and height $L_y = 32\sigma$ is shaken by a sinusoidal vibration with frequency f (period $\tau_{box} = 1/f$) and amplitude 2.6σ. Collisions with the elastic walls inject energy and allow the system to reach a stationary state. We have checked that possible inelastic collisions with the walls hardly affect the results. Gravity – set to $g = -1.7\sigma f^2$ in order to be consistent with the experiment – has a negligible influence on the measured quantities. We have varied the restitution coefficient from 0.8 up to 0.99 (glass beads yield on average $\alpha \approx 0.9$) and the total area coverage from 0.138 (i.e. $N = 270$) up to 0.32 ($N = 620$). In Fig. 17-left a snapshot of the system is shown. During the simulations the main physical observables are statistically stationary. The local area coverage field $\Phi(x, y)$ and the granular temperature field $T(x, y)$ (defined in two dimensions as the local average kinetic energy per particle) are almost uniform in the horizontal direction, apart from small layers near the side walls. In Fig. 17-right the profiles $\Phi(y) = (1/L_x) \int dx \Phi(x, y)$ and $T(y) = (1/L_x) \int dx T(x, y)$ are shown to be symmetric with respect to the bottom and the top of the box. Following the experimental procedure, we have focused our attention on a "window" in the center of the box, fixed in the laboratory frame, of width $2L_x/5$ and height $L_y/3$, marked in Fig. 17-left. Apart from the negligible change of potential energy due to gravity, the total kinetic energy of the particles inside the window, changes during a time τ because of two contributions: $\Delta K_\tau = Q_\tau - I_\tau$ where Q_τ is the kinetic energy transported by particles through the boundary of the window (summed when going-in and subtracted when going-out) and I_τ

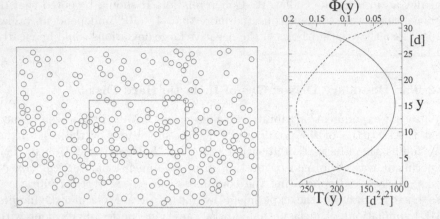

Fig. 17. *Left*: Snapshot of the system considered for MD simulations, with the inner region marked by the solid rectangle. *Right*: Corresponding vertical profiles of density ($\Phi(y)$, *dashed line*) and temperature ($T(y)$, *solid line*). The *dotted lines* mark the bottom and top boundaries of the inner region. Here $N = 270$ and $\alpha = 0.9$. The mean free path is $\sim 5.7d$

is the kinetic energy dissipated in inelastic collisions during time τ. For several values of τ we have measured, as in the experiments, Q_τ which is related to the kinetic contribution to the heat flux (we checked that inclusion of the collisional contribution, even if non small [HMZ04], does not change the picture). With $N = 270$ and $\alpha = 0.9$ the characteristic times are the mean free time $\tau_{col} \approx 0.47\tau_{box}$, the diffusion time across the window $\tau_{diff} = 0.82\tau_{box}$ and the mean time between two subsequent crossings of particles (from outside to inside) $\tau_{cross} \approx 0.039\tau_{box}$.

We define the injected power as $q_\tau = Q_\tau/\tau$ and two relevant probability density functions (pdfs): $f_Q(Q_\tau)$ and $f_q(q_\tau)$. Figure 18a shows $f_q(q_\tau)$ for different values of τ. A direct comparison with Fig. 3 of [FM04] suggests a fair agreement between simulations of inelastic hard disks and the experiment. The pdfs are strongly non-Gaussian and asymmetric, becoming narrower as τ is increased. At small τ a strong peak in $q_\tau = 0$ is visible. More interestingly, $f_q(q_\tau)$ at small values of τ has two different exponential tails, i.e. $f_q(q_\tau) \sim \exp(\mp\beta_\pm\tau q_\tau)$ when $q_\tau \to \pm\infty$ with $\beta_- > \beta_+$. The peak and the exponential tails at small τ are observed also in the experiment (see Fig. 3 of [FM04]) and in similar simulations [AFMP01]. In Fig. 18b we display $\log[f_q(q_\tau)/f_q(-q_\tau)]/\tau$ versus q_τ, which is equivalent to the graph of $\pi_\tau(q_\tau) - \pi_\tau(-q_\tau)$ versus q_τ where $\pi_\tau(q_\tau) = \log[f_Q(\tau q_\tau)]/\tau$. From Fig. 18 it appears that at large values of τ, $\pi_\tau(q_\tau) - \pi_\tau(-q_\tau)$ is linear with a τ-independent slope $\beta_{eff} \neq 1$. We have measured β_{eff} with various choices of the restitution coefficient α and of the covered area fraction finding similar results. Feitosa and Menon [FM04] report $\beta_{eff}T_{gran} \sim 0.25$ where T_{gran} is

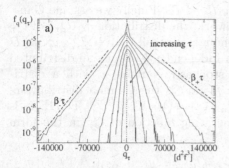 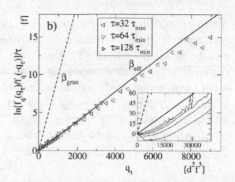

Fig. 18. (a) Pdfs of injected power $f_q(q_\tau)$ from MD simulations for different values of $\tau = (1, 2, 4, 8, 16, 32) \times \tau_{min}$ with $\tau_{min} = 0.015\tau_{box}$. Here $N = 270$ and $\alpha = 0.9$. The distributions are shifted vertically for clarity. The *dashed lines* put in evidence the exponential tails of the pdf at $\tau = \tau_{min}$. (b) Plot of $(1/\tau) \log[f_q(q_\tau)/f_q(-q_\tau)]$ versus q_τ from MD simulations (same parameters as above) at large values of τ. The *solid curve* is a linear fit (with slope β_{eff}) of the data at $\tau = 128\tau_{min}$. The *dashed line* has a slope $\beta_{gran} = 1/T_{gran}$. In the inset the same graph is shown for small values of $\tau = (1, 2, 4, 8) \times \tau_{min}$ (from *bottom to top*)

the mean granular temperature in the observation window. Similar values are measured in our MD simulations. At area fraction 13.8% and $\alpha = 0.9$ we have $\beta_{eff}T_{gran} \approx 0.23$. At fixed α and increasing area fraction, $\beta_{eff}T_{gran}$ slightly increases, as in the experiment. As $\alpha \rightarrow 1$ the slope β_{eff} decreases. At $\alpha = 1$ (without gravity and external driving) the distribution of Q_τ is symmetrical and $\beta_{eff} = 0$, indicating that $1/\beta_{eff}$ is not a physically relevant temperature concept. Interestingly, it appears that β_{eff} is a non hydrodynamic quantity: different systems may show the *same* density and temperature profiles, with very different values of β_{eff}.

We now adopt a coarse-grained description of the experiment which is able to entirely capture the observed phenomenology. The measured flow of energy is given by

$$Q_\tau = \frac{1}{2} \left(\sum_{i=1}^{n_+} v_{i+}^2 - \sum_{i=1}^{n_-} v_{i-}^2 \right), \tag{96}$$

where n_- (n_+) is the number of particles leaving (entering) the window during the time τ, and v_{i-}^2 (v_{i+}^2) are the squared moduli of their velocities. In order to analyze the statistics of Q_τ we take n_- and n_+ as Poisson-distributed random variables with average $\omega\tau$, where ω corresponds to the inverse of the crossing time τ_{cross}. In doing so we neglect correlations among particles entering or leaving successively the central region. A key point, supported by direct observation in the numerical experiment, lies in the assumption that the velocities $\mathbf{v}_{i+}$ and $\mathbf{v}_{i-}$ come from populations with different temperatures T_+ and T_- respectively. Indeed, compared with the population entering the central region, those particles that leave it have suffered on average more inelastic collisions, so that $T_- < T_+$. Finally we assume Gaussian velocity pdfs. Within such a framework, the distribution $f_Q(Q_\tau)$ of Q_τ can be studied analytically. Here it is enough to recall that $\frac{1}{2}\sum_{i=1}^n v_i^2$, in D dimensions, if each component of $\mathbf{v}_i$ is independently Gaussian-distributed with zero mean and variance T, is a stochastic variable x with a distribution $\chi_{n,T}(x) = f_{1/T,Dn/2}(x)$, where $f_{\alpha,\nu}(x)$ is the Gamma distribution, and whose generating function reads $\tilde\chi_{n,T}(z) = (1 - Tz)^{-Dn/2}$ [Fel71]. It is then straightforward to obtain the generating function of Q_τ in the form $\tilde f_Q(z) = \exp[\tau\mu(z)]$ with

$$\mu(z) = \omega\left(-2 + (1 - T_+z)^{-D/2} + (1 + T_-z)^{-D/2}\right). \tag{97}$$

We observe that $\tilde f_Q(z)$ has two poles in $z = \pm 1/T_\pm$ and two branch cuts on the real axis for $z > 1/T_+$ and $z < -1/T_-$. From $\mu(z)$ we immediately obtain the cumulants of $f_Q(Q_\tau)$ through the formula $\langle Q^n\rangle_c = \tau\frac{d^n}{dz^n}\mu(0)$.

For $\tau \to \infty$ the large deviation theory states that $f_Q(Q_\tau) \sim \exp(\tau\pi_\infty(Q_\tau/\tau))$ and $\pi_\infty(q)$ can be obtained from $\mu(z)$ through a Legendre transform, i.e. $\pi_\infty(q) = max(\mu(z) - qz)$. The study of the singularities of $\mu(z)$ reveals the behavior of the large deviation function $\pi_\infty(q)$ for $q \to \pm\infty$. It can be seen that

$$\pi_\infty(q) \sim -\frac{q}{T_+} \ (q \to \infty), \quad \pi_\infty(q) \sim \frac{q}{T_-} \ (q \to -\infty). \tag{98}$$

We emphasize however that it is almost impossible to appreciate these tails in simulations and in experiments, since the statistics for large values of q and τ is very poor.

A Gallavotti–Cohen-type relation [GC95, LS99, Kur98], e.g. $\pi_\infty(q) - \pi_\infty(-q) = \beta q$ for any q and an arbitrary value of β would imply $\mu(z) = \mu(\beta - z)$. One can see that such a β does not exist, i.e. the fluctuations of Q_τ do not satisfy a Gallavotti–Cohen-like relation. The observed linearity of the graph $\log[f_q(q_\tau)/f_q(-q_\tau)]/\tau = \pi(q_\tau) - \pi(-q_\tau)$ versus q_τ can be explained by the same observation pointed out in the previous subsection: at large values of τ it is extremely difficult, in simulations as well as in experiments, to reach large values of q, while for small q, $\pi(q) - \pi(-q) \approx 2\pi'(0)q + o(q^3)$, i.e. a straight line with a slope $\beta_{eff} = 2\pi'(0)$ is likely to be observed. It has been already shown [Far02] that in dissipative systems deviations from the FR can be hidden by insufficient statistics at high values of q. The knowledge of $\mu(z)$ is useful to predict this slope. At large τ, $\pi'(0) \approx \Pi'(0) = -z^*(0)$ where $z^*(q)$ is the value of z for which $\mu(z) - qz$ is extremal. This gives

$$\beta_{eff} = \frac{T_+^\delta - T_-^\delta}{T_+^{\delta+1} + T_-^{\delta+1}} \quad \text{with} \quad \delta = \frac{2}{2 + D}. \tag{99}$$

When $T_+ = T_-$ (i.e. if $\alpha = 1$) $\beta_{eff} = 0$. We emphasize that β_{eff} does not depend upon ω. We have compared with success these predictions with the numerical and experimental results, measuring the temperatures T_+ and T_- in the simulation.

What happens for small values of τ? We note that $\tilde{f}_Q(z)$ has the form $\exp(\tau\mu(z))$ for *any* value of τ and not only for large τ. Therefore at small τ one can expand the exponential, obtaining $\tilde{f}_Q(z) \sim 1 + \omega\tau(-2 + (1 - T_+z)^{-D/2} + (1 + T_-z)^{-D/2})$. This immediately leads to an analytical expression for $f_Q(Q_\tau) = \text{const} \times \delta(Q_\tau) + \chi_{1,T_+}(Q_\tau) + \chi_{1,T_-}(-Q_\tau)$, which fairly accounts for the strong peak which is observed in the experiment and in the simulations, and predicts exponential tails for $f_Q(Q_\tau)$: $\chi_{1,T}(x) \propto x^{D/2-1}\exp(-x/T)$ so that $\beta_+ = 1/T_+$ and $\beta_- = 1/T_-$. This suggests an experimental test of this theoretical approach: the measure at small values of τ of the slopes of the exponential tails of $f_Q(Q_\tau)$ should coincide with a direct measure of T_+ and T_-. However, we point out that the values of β_+ and β_- obtained by fitting the tails in the hard disks simulation, using values as small as $\tau = 0.00015\tau_{box}$ yield estimates of T_+ and T_- which are larger (by a factor ~ 1.6) than those found by a direct measure. This disagreement brings the limits of such a simple two-temperature picture to the fore. In the simulation and in the original experiment the measured injected energy is indeed the sum of several different contributions, namely $Q_\tau \approx Q_\tau^{xx} + Q_\tau^{xy} + Q_\tau^{yx} + Q_\tau^{yy}$ where Q_τ^{ij} is the kinetic energy transported by the i component of the velocity by particles crossing the boundary through a wall perpendicular to direction j. Two main differences with the simplified interpretation given above arise: (a) there are *two* couples of temperatures, i.e. T_+^x, T_-^x as well as T_+^y, T_-^y [BC98, ML98, BK03]; (b) the diagonal contributions Q_τ^{jj} are sums of squares of velocities whose distribution is not a Gaussian but is $\sim v\exp(-v^2/T)$, since the probability of crossing is biased by the velocity itself. The calculation of $f_Q(Q_\tau)$ is still feasible, with qualitatively similar results.

6 The Dynamics of a Tracer Particle as a Non-Equilibrium Markov Process

In the search for a quantity that is, more rigorously, related to the "entropy production" in a granular gas, we consider in this section the projection of the dynamics of the gas onto that of a tracer particle, which is easier since it is equivalent to a jump Markov process. We are interested in the dynamics of a tracer granular particle in a homogeneous and dilute gas of grains which is driven by an unspecified energy source. The requirements are that the gas is dilute, spatially homogeneous and time translational invariant. The gas is characterized by its velocity probability density function $P(\mathbf{v})$ which, for the sake of simplicity, will be considered of the form $P(\mathbf{v}) = \frac{1}{(2\pi T)^{d/2}}\exp\left(-\frac{v^2}{2T}\right)(1 + a_2 S_2^d(v^2/2T))$, where S_2^d is the d-dimensional second Sonine polynomial already defined in (2.2). The gas is therefore parametrized by its temperature T and its second Sonine coefficient a_2 which measures its non-Gaussianity.

The linear Boltzmann equation for the tracer, in generic dimension d, reads:

$$\frac{dP_*(\mathbf{v},t)}{dt} = \frac{1}{\ell} \int d\mathbf{v}_1 \int d\mathbf{v}_2 \int' d\hat{\omega} |(\mathbf{v}_1 - \mathbf{v}_2) \cdot \hat{\omega}| P_*(\mathbf{v}_1) P(\mathbf{v}_2) \times$$

$$\times \left\{ \delta\left(\mathbf{v} - \mathbf{v}_1 + \frac{1+\alpha}{2}[(\mathbf{v}_1 - \mathbf{v}_2) \cdot \hat{\omega}]\hat{\omega}\right) - \delta(\mathbf{v} - \mathbf{v}_1) \right\} \quad (100)$$

where $P_*(v)$ is the velocity pdf of the test particle and the primed integral again indicates that the integration is performed on all angles that satisfy $(\mathbf{v}_1 - \mathbf{v}_2) \cdot \hat{\omega} > 0$. The mean free path ℓ appears in front of the collision integrals. In the following (when not stated differently) we will put $\ell = 1$, which can be always obtained by a rescaling of time.

We rewrite (100) as a Master equation for a Markov jump process [PVTvW06]:

$$\frac{dP_*(\mathbf{v},t)}{dt} = \int d\mathbf{v}_1 P_*(\mathbf{v}_1) K(\mathbf{v}_1, \mathbf{v}) - \int d\mathbf{v}_1 P_*(\mathbf{v}) K(\mathbf{v}, \mathbf{v}_1). \quad (101)$$

The transition rate $K(\mathbf{v}, \mathbf{v}')$ of jumping from $\mathbf{v}$ to $\mathbf{v}'$ is given by the following formula:

$$K(\mathbf{v}, \mathbf{v}') = \left(\frac{2}{1+\alpha}\right)^2 \frac{1}{\ell} |\Delta \mathbf{v}|^{2-d} \int d\dot{\mathbf{v}}_{2\tau} P[\mathbf{v}_2(\mathbf{v}, \mathbf{v}', \mathbf{v}_{2\tau})], \quad (102)$$

where $\Delta \mathbf{v} = \mathbf{v}' - \mathbf{v}$ denotes the change of velocity of the test particle after a collision. The vectorial function $\mathbf{v}_2$ is defined as

$$\mathbf{v}_2(\mathbf{v}, \mathbf{v}', \mathbf{v}_{2\tau}) = v_{2\sigma}(\mathbf{v}, \mathbf{v}')\hat{\boldsymbol{\sigma}}(\mathbf{v}, \mathbf{v}') + \mathbf{v}_{2\tau}, \quad (103)$$

where $\hat{\boldsymbol{\sigma}}(\mathbf{v}, \mathbf{v}')$ is the unitary vector parallel to $\Delta \mathbf{v}$, while $\mathbf{v}_{2\tau}$ is entirely contained in the $(d-1)$-dimensional space perpendicular to $\Delta \mathbf{v}$ (i.e. $\mathbf{v}_{2\tau} \cdot \Delta \mathbf{v} = 0$). This implies that the integral in expression (102) is $(d-1)$-dimensional. Finally, to fully determine the transition rate (102), the expression of $v_{2\sigma}$ is needed:

$$v_{2\sigma}(\mathbf{v}, \mathbf{v}') = \frac{2}{1+\alpha} |\Delta \mathbf{v}| + \mathbf{v} \cdot \hat{\boldsymbol{\sigma}}. \quad (104)$$

6.1 Detailed Balance

Here, we obtain a simple expression for the ratio between $K(\mathbf{v}, \mathbf{v}')$ and $K(\mathbf{v}', \mathbf{v})$. When exchanging $\mathbf{v}$ with $\mathbf{v}'$ the unitary vector $\hat{\boldsymbol{\sigma}}$ changes sign. Furthermore one has that $v_{2\sigma}(\mathbf{v}, \mathbf{v}') \neq v_{2\sigma}(\mathbf{v}', \mathbf{v})$. From all these considerations and from (102) one obtains immediately:

$$\frac{K(\mathbf{v}, \mathbf{v}')}{K(\mathbf{v}', \mathbf{v})} = \frac{\int d\mathbf{v}_{2\tau} P[\mathbf{v}_2(\mathbf{v}, \mathbf{v}')]}{\int d\mathbf{v}_{2\tau} P[\mathbf{v}_2(\mathbf{v}', \mathbf{v})]} \equiv \frac{P[v_{2\sigma}(\mathbf{v}, \mathbf{v}')]}{P[v_{2\sigma}(\mathbf{v}', \mathbf{v})]}. \tag{105}$$

We note that this ratio depends only on the choice of the pdf of the gas, P, and not on the other parameters (such as α). However in realistic situations (experiments or Molecular Dynamics simulations) P is not a free parameter but is determined by the choice of the setup (e.g. external driving, material details, geometry of the container, etc.).

Introducing the short-hand notation $v_{2\sigma} = v_{2\sigma}(\mathbf{v}, \mathbf{v}')$, $v'_{2\sigma} = v_{2\sigma}(\mathbf{v}', \mathbf{v})$ and $v_{\sigma}^{(\prime)} = \mathbf{v}^{(\prime)} \cdot \hat{\boldsymbol{\sigma}}$, we also note that

$$(v'_{2\sigma})^2 = v_{2\sigma}^2 + (v_\sigma + v'_\sigma)^2 - 2v_{2\sigma}(v_\sigma + v'_\sigma), \tag{106}$$

from which it follows that

$$\Delta_2 = (v_{2\sigma})^2 - (v'_{2\sigma})^2 = -\Delta - 2\frac{1-\alpha}{1+\alpha}\Delta = -\frac{3-\alpha}{1+\alpha}\Delta, \tag{107}$$

where $\Delta = v_\sigma^2 - (v'_\sigma)^2 \equiv |v|^2 - |v'|^2$, i.e. the kinetic energy lost by the test-particle during one collision. When $\alpha = 1$ then $\Delta_2 = -\Delta$ (energy conservation). From the above considerations it follows that

- in the *Gaussian* case, it is found

$$\log \frac{K(\mathbf{v}, \mathbf{v}')}{K(\mathbf{v}', \mathbf{v})} = \frac{\Delta}{2T} + 2\frac{1-\alpha}{1+\alpha}\frac{\Delta}{2T} = \frac{3-\alpha}{1+\alpha}\frac{\Delta}{2T} \tag{108}$$

- in the *First Sonine Correction* case, it is found

$$\log \frac{K(\mathbf{v}, \mathbf{v}')}{K(\mathbf{v}', \mathbf{v})} = \frac{3-\alpha}{1+\alpha}\frac{\Delta}{2T} + \log \frac{\left\{1 + a_2 S_2^{d=1}\left[\frac{\left(\frac{2}{1+\alpha}(v'_\sigma - v_\sigma) + v_\sigma\right)^2}{2T}\right]\right\}}{\left\{1 + a_2 S_2^{d=1}\left[\frac{\left(\frac{2}{1+\alpha}(v_\sigma - v'_\sigma) + v'_\sigma\right)^2}{2T}\right]\right\}} \tag{109}$$

In the case where $P(v)$ is a Gaussian with temperature T, it is immediate to observe that

$$P_*(\mathbf{v})K(\mathbf{v}, \mathbf{v}') = P_*(\mathbf{v}')K(\mathbf{v}', \mathbf{v}) \tag{110}$$

if P_* is equal to a Gaussian with temperature $T' = \frac{\alpha+1}{3-\alpha}T \leq T$. This means that there is a Gaussian stationary solution of (101) (in the Gaussian-bulk case), which satisfies detailed balance. The fact that such a Gaussian with a different temperature T' is an exact stationary solution was known from [MP99]. It thus turns out that detailed balance is satisfied, even out of thermal equilibrium. Of course this is an artifact of such a model: it is highly unrealistic that a granular gas yields a Gaussian velocity pdf. As soon as the gas velocity pdf $P(v)$ ceases to be Gaussian, detailed balance is violated, i.e. the stationary process performed by the tracer particle is no more in equilibrium within the thermostatting gas. We will see in Sect. 6.2 how to characterize this departure from equilibrium.

6.2 Action Functionals

From the previous section we have learnt that the dynamics of the velocity of a tracer particle immersed in a granular gas is equivalent to a Markov process with well defined transition rates. This means that the velocity of the tracer particle stays in a state $\mathbf{v}$ for a random time $t \geq 0$ distributed with the law $r(\mathbf{v})e^{-r(\mathbf{v})t}dt$ and then makes a transition to a new value $\mathbf{v}'$ with a probability $r(\mathbf{v})^{-1}K(\mathbf{v}, \mathbf{v}')$, with $r(\mathbf{v}) = \int d\mathbf{v}' K(\mathbf{v}, \mathbf{v}')$. At this point it is interesting to ask about some characterization of the non-equilibrium dynamics, i.e. of the violation of detailed balance, which we know to happen whenever the surrounding granular gas has a non-Gaussian distribution of velocity.

To this extent, we define two different action functionals, following [LS99]:

$$W(t) = \sum_{i=1}^{n(t)} \log \frac{K(\mathbf{v}_i \to \mathbf{v}'_i)}{K(\mathbf{v}'_i \to \mathbf{v}_i)} \tag{111a}$$

$$\overline{W}(t) = \log \frac{P_*(\mathbf{v}_1)}{P_*(\mathbf{v}'_{n(t)})} + \sum_{i=1}^{n(t)} \log \frac{K(\mathbf{v}_i \to \mathbf{v}'_i)}{K(\mathbf{v}'_i \to \mathbf{v}_i)} \tag{111b}$$

$$\equiv \log \frac{\mathcal{P}(\mathbf{v}_1 \to \mathbf{v}_2 \to ... \to \mathbf{v}_{n(t)})}{\mathcal{P}(\mathbf{v}_{n(t)} \to \mathbf{v}_{n(t)-1} \to ... \to \mathbf{v}_1)} \tag{111c}$$

where i is the index of collision suffered by the tagged particle, $\mathbf{v}_i$ is the velocity of the particle before the i-th collision, $\mathbf{v}'_i$ is its post-collisional velocity, $n(t)$ is the total number of collisions in the trajectory from time 0 up to time t, and K is the transition rate of the jump due to the collision. Finally, we have used the notation $\mathcal{P}(\mathbf{v}_1 \to \mathbf{v}_2 \to ... \to \mathbf{v}_n)$ to identify the probability of observing the trajectory $\mathbf{v}_1 \to \mathbf{v}_2 \to ... \to \mathbf{v}_n$. The quantities $W(t)$ and $\overline{W}(t)$ are different for each different trajectory (i.e. sequence of jumps) of the tagged particle. Note that the first term $\log \frac{P_*(\mathbf{v}_1)}{P_*(\mathbf{v}'_{n(t)})}$ in the definition of $\overline{W}(t)$, (111c), is non-extensive in time. The two above functionals have the following properties:

- $W(t) \equiv 0$ if there is exact symmetry, i.e. if $K(\mathbf{v}_i \to \mathbf{v}_{i+1}) = K(\mathbf{v}_{i+1} \to \mathbf{v}_i)$ (e.g. in the microcanonical ensemble); $\overline{W}(t) \equiv 0$ if there is detailed balance (e.g. any equilibrium ensemble).
- We expect that, for large enough t, for almost all the trajectories $\lim_{s \to \infty} W(s)/s = \lim_{s \to \infty} \overline{W}(s)/s = \langle W(t)/t \rangle = \langle \overline{W}(t)/t \rangle$; here (since the system under investigation is ergodic and stationary) the meaning of $\langle \rangle$ is intuitively an average over many independent segments of a single very long trajectory.
- For large enough t: (1) at equilibrium $\langle W(t) \rangle = \langle \overline{W}(t) \rangle = 0$; (2) out of equilibrium (i.e. if detailed balance is not satisfied) those two averages are positive; we use those equivalent averages, at large t, to characterize the distance from equilibrium of the stationary system.

- If $S(t) = - \int \mathrm{d}v P_*(v,t) \log P_*(v,t)$ is the entropy associated to the pdf of the velocity of the tagged particle $P_*(v,t)$ at time t (e.g. $-H$ where H is the Boltzmann-H function), then

$$\frac{\mathrm{d}}{\mathrm{d}t}S(t) = R(t) - A(t) \qquad (112)$$

where $R(t)$ is always non-negative, $A(t)$ is linear with respect to P_* and, finally, $\langle W(t) \rangle \equiv \int_0^t \mathrm{d}t' A(t')$. This leads to consider $W(t)$ equivalent to the contribution of a single trajectory to the total entropy flux. In a stationary state $A(t) = R(t)$ and therefore the flux is equivalent to the production; this property has been recognized in [LS99].

- FR_W (Lebowitz–Spohn–Gallavotti–Cohen fluctuation relation): $\pi(w) - \pi(-w) = w$ where $\pi(w) = \lim_{t \to \infty} \frac{1}{t} \log f_W^t(tw)$ and $f_W^t(x)$ is the probability density function of finding $W(t) = x$ at time t; at equilibrium the FR_W has no content; note that in principle $\pi'(w,t) = \frac{1}{t} \log f_W^t(tw) \neq \pi(w)$ at any finite time; a generic derivation of this property has been obtained in [LS99], while a rigorous proof with more restrictive hypothesis is in [Mae99]; the discussion for the case of a Langevin equation is in [Kur98].

- $FR_{\overline{W}}$ (Evans–Scarles fluctuation relation): $\overline{\pi}(w,t) - \overline{\pi}(-w,t) = w$ where $\overline{\pi}(w,t) = \frac{1}{t} \log f_{\overline{W}}^t(tw)$ and $f_{\overline{W}}^t(x)$ is the probability density function of finding $\overline{W}(t) = x$ at time t; at equilibrium the $FR_{\overline{W}}$ has no content; this relation is derived in [LS99]; the analogy between this relation and the Evans–Searles fluctuation relation [ES94, ES02] has been put forward in [PVTvW06].

A detailed numerical study [PVTvW06] of the fluctuations of $W(t)$ and $\overline{W}(t)$ in this model has shown on the one hand that, out of equilibrium (i.e. when the surrounding gas is non-Gaussian), the $FR_{\overline{W}}$ is always satisfied. On the other hand the FR_W is always violated, even if it was expected on the basis of the arguments given in [LS99]. The difference between the two functionals defined in (111) is a term which is non-extensive in time, but which has fluctuations whose distribution has exponential tails and therefore, in principle, can contribute to the large deviation function of $W(t)$ [PRV06, Vis06]. Such a failure of a large time Fluctuation Relation, which is much more pronounced in the near-to-equilibrium cases, is similar to that observed in other systems [ESR05, Far02, vZC03, BGGZ05].

7 Conclusions

The study of the fluctuations of global physical quantities in a granular gas is at its very beginning. In the lack of a general rigorous theory in the framework of non equilibrium statistical mechanics, experiments and numerical simulations are the main source of results, together with few exact analytical calculations. In this review of recent results [VPB$^+$05, PVB$^+$05,

VPB⁺06a, PVTvW06, VPB⁺06b] we have indicated some routes that have been followed, focusing on two global quantities (total energy and energy injection rate) that are of interest in nowadays physics of non-equilibrium systems [BHP98, BdSMRM05, AFMP01, AFFM04, Far02, Far04]. On one hand we have shown that total energy fluctuations have a pdf that strongly depends on the model considered. We have also pointed out that definitive inferences about the presence of correlations, starting from the observation of "anomalous" pdfs of total energy, must be drawn with caution, since the lack of spatial or temporal translational invariance may play a major role. On the other hand we have presented a method to calculate the large deviation function of injected power in a granular gas: this method strongly suggests the disappearance of a negative branch in such large deviation function. This result is a direct consequence of the time-irreversibility of inelastic collisions: injected power fluctuations are dominated at large times by the energy dissipated in collisions, which is always positive. Finally we have sketched a recipe to obtain a quantity related to time-reversal asymmetry (i.e. violation of detailed balance) whose large deviations can be both positive and negative. This quantity has the advantage of being measurable in experiments, but the disadvantage of not having an obvious "macroscopic" counterpart. It contains in fact information on the non-Gaussianity of the velocity pdf of the gas. This consideration is in our opinion the main open issue in this study.

References

[AFFM04]   S. Aumaître, J. Farago, S. Fauve, and S. McNamara. Energy and poweer fluctuations in vibrated granular gases. *Eur. Phys. J. B*, 42:255, 2004.

[AFMP01]   S. Aumaître, S. Fauve, S. McNamara, and P. Poggi. Power injected in dissipative systems and the fluctuation theorem. *Eur. Phys. J. B*, 19:449–460, 2001.

[BC98]   J. J. Brey and D. Cubero. Steady state of a fluidized granular medium between two walls at thesame temperature. *Phys. Rev. E*, 57(2):2019–2029, 1998.

[BC01]   J. J. Brey and D. Cubero. Hydrodynamic transport coefficients of granular gases. In T. Pöschel and S. Luding, editors, *Granular Gases*, pages 59–79, Berlin, 2001. Springer.

[BCG00]   A. V. Bobylev, J. A. Carrillo, and I. M. Gamba. On some properties of kinetic and hydrodynamic equations for inelastic interactions. *J. Stat. Phys.*, 98:743, 2000.

[BDKS98]   J. J. Brey, J. W. Dufty, C. S. Kim, and A. Santos. Hydrodynamics for granular flow at low density. *Phys. Rev. E*, 58(4):4638, 1998.

[BdSMRM04]   J. J. Brey, M. I. G. de Soria, P. Maynar, and M. J. Ruiz-Montero. Energy fluctuations in the homogeneous cooling state of granular gases. *Phys. Rev. E*, 70(011302), 2004.

[BdSMRM05]   J. J. Brey, M. I. G. de Soria, P. Maynar, and M. J. Ruiz-Montero. Scaling and universality of critical fluctuations in granular gases. *Phys. Rev. Lett.*, 94:098001, 2005.

[Ber05]   E. Bertin. Global fluctuations and gumbel statistics. cond-mat/
            0506166, to appear on Phys. Rev. Lett., 2005.
[BGGZ05]   F. Bonetto, G. Gallavotti, A. Giuliani, and F. Zamponi.
            Chaotic hypothesis, fluctuation theorem and singularities. cond-
            mat/0507672, 2005.
[BHP98]   S. T. Bramwell, P. C. W. Holdsworth, and J.-F. Pinton. Universality
            of rare fluctuations in turbulence and critical phenomena. *Nature*,
            396:552–554, 1998.
[Bir94]   G. A. Bird. *Molecular Gas Dynamics and the Direct Simulation of
            Gas Flows*. Clarendon, Oxford, 1994.
[BK03]   D. L. Blair and A. Kudrolli. Collision statistics of driven granular
            materials. *Phys. Rev. E*, 67:041301, 2003.
[BMP02]   A. Baldassarri, U. Marini Bettolo Marconi, and A. Puglisi. Influ-
            ence of correlations on the velocity statistics of scalar granular gases.
            Europhys. Lett., 58:14–20, 2002. cond-mat/0111066.
[BNK02]   E. Ben-Naim and P. L. Krapivsky. Nontrivial velocity distributions
            in inelastic gases. *J. Phys. A: Math. Gen.*, 35:L147–L152, 2002.
            cond-mat/0111044.
[BNK03]   E. Ben-Naim and P. L. Krapivsky. The inelastic maxwell model.
            In *Lecture Notes in Physics*, volume 624, page 65, Berlin, 2003.
            Springer. cond-mat/0301238.
[BRMM00]   J. Javier Brey, M. J. Ruiz-Montero, and F. Moreno. Boundary con-
            ditions and normal state for a vibrated granular fluid. *Phys. Rev. E*,
            62:5339–5346, 2000.
[BT02]   A. Barrat and E. Trizac. Molecular dynamics simulations of vibrated
            granular gases. *Phys. Rev. E*, 66(5):051303, 2002. cond-mat/0207267.
[BTE05]   A. Barrat, E. Trizac, and M. H. Ernst. Granular gases: dynamics
            and collective effects. *J. Phys. Condens. Matter*, 17:S2429, 2005.
[CC60]   S. Chapman and T. G. Cowling. *The mathematical theory of
            nonuniform gases*. Cambridge University Press, London, 1960.
[CDPT03]   F. Coppex, M. Droz, J. Piasecki, and E. Trizac. On the first sonine
            correction for granular gases. *Physica A*, 329:114, 2003.
[CPM07]   G. Costantini, A. Puglisi, and U. Marini Bettolo Marconi. Velocity
            fluctuations in a one dimensional inelastic Maxwell model. *J. Stat.
            Mech.*, page P08031, 2007.
[dGM69]   S. R. de Groot and P. Mazur. *Non-equilibrium thermodynamics*.
            North-Holland, Amsterdam, 1969.
[EB02]   M. H. Ernst and R. Brito. High-energy tails for inelastic maxwell
            models. *Europhys. Lett.*, 58:182, 2002.
[EC81]   M. H. Ernst and E. G. D. Cohen. Nonequilibrium fluctuations in
            μ-space. *J. Stat. Phys*, 25(1):153, 1981.
[ECM93]   D. J. Evans, E. G. D. Cohen, and G. P. Morriss. Probability of second
            law violations in shearing steady states. *Phys. Rev. Lett.*, 71:2401,
            1993.
[ES94]   D. J. Evans and D. J. Searles. Equilibrium microstates which gen-
            erate second law violating steady states. *Phys. Rev. E*, 50:1645,
            1994.
[ES02]   D. J. Evans and D. J. Searles. The fluctuation theorem. *Adv. Phys.*,
            51:1529, 2002.

[ESR05]   D. J. Evans, D. J. Searles, and L. Rondoni. Application of the gallavotti-cohen fluctuation relation to thermostated steady states near equilibrium. *Phys. Rev. E*, 71:056120, 2005.

[Far02]   J. Farago. Injected power fluctuations in langevin equation. *J. Stat. Phys.*, 107:781, 2002.

[Far04]   J. Farago. Power fluctuations in stochastic models of dissipative systems. *Physica A*, 331:69–89, 2004.

[Fel71]   W. Feller. *Probability Theory and its Applications*. John Wiley & Sons, New York, 1971.

[FM04]   K. Feitosa and N. Menon. Fluidized granular medium as an instance of the fluctuation theorem. *Phys. Rev. Lett*, 92:164301, 2004.

[GC95]   G. Gallavotti and E. G. D. Cohen. Dynamical ensembles in nonequilibrium statistical mechanics. *Phys. Rev. Lett.*, 74:2694, 1995.

[GZBN97]   E. L. Grossman, T. Zhou, and E. Ben-Naim. Towards granular hydrodynamics in two-dimensions. *Phys. Rev. E*, 55:4200, 1997.

[HBB00]   C. Henrique, G. Batrouni, and D. Bideau. Diffusion as mixing mechanism in granular materials. *Phys. Rev. E*, 63:011304, 2000. cond-mat/0003354.

[HMZ04]   O. Herbst, P. M. Müller, and A. Zippelius. Local heat flux and energy loss in a 2d vibrated granular gas. cond-mat/0412334, 2004.

[Kum98]   V. Kumaran. Temperature of a granular material "fluidized" by external vibrations. *Phys. Rev. E*, 57(5):5660–5664, 1998.

[Kur98]   J. Kurchan. Fluctuation theorem for stochastic dynamics. *J. Phys. A.*, 31:3719, 1998.

[LS99]   J. L. Lebowitz and H. Spohn. A gallavotti-cohen-type symmetry in the large deviation functional for stochastic dynamics. *J. Stat. Phys.*, 95:333, 1999.

[Mae99]   C. Maes. The fluctuation theorem as a gibbs property. *J. Stat. Phys.*, 95:367, 1999.

[Max67]   J. C. Maxwell. On the dynamical theory of gases. *Phil. Trans.*, 157: 49, 1867.

[MB97]   S. McNamara and J.-L. Barrat. The energy flux into a fluidized granular medium at a vibrating wall. *Phys. Rev. E*, 55:7767, 1997.

[ML98]   S. McNamara and S. Luding. Energy flows in vibrated granular media. *Phys. Rev. E*, 58:813–822, 1998.

[MP99]   P. A. Martin and J. Piasecki. Thermalization of a particle by dissipative collisions. *Europhys. Lett.*, 46(5):613, 1999.

[MS00]   José Maria Montanero and Andrés Santos. Computer simulation of uniformly heated granular fluids. *Granular Matter*, 2(2):53–64, 2000. cond-mat/0002323.

[MSS01]   S. J. Moon, M. D. Shattuck, and J. B. Swift. Velocity distribution and correlations in homogeneously heated granular media. *Phys. Rev. E*, 64:031303, 2001. cond-mat/0105322.

[PBL02]   A. Puglisi, A. Baldassarri, and V. Loreto. Fluctuation-dissipation relations in driven granular gases. *Phys. Rev. E*, 66:061305, 2002. cond-mat/0206155.

[PBV07]   A. Puglisi, A. Baldassarri, and A. Vulpiani. Violation of the Einstein relation in granular fluids: the role of correlations. *J. Stat. Mech.*, page P08016, 2007.

[PLM+98]   A. Puglisi, V. Loreto, U. M. B. Marconi, A. Petri, and A. Vulpiani. Clustering and non-gaussian behavior in granular matter. *Phys. Rev. Lett.*, 81:3848, 1998.

[PO98]   G. Peng and T. Ohta. Steady state properties of a driven granular medium. *Phys. Rev. E*, 58:4737–46, 1998. cond-mat/9710119.

[PRV06]   A. Puglisi, L. Rondoni, and A. Vulpiani. Relevance of initial and final conditions for the fluctuation relation in Markov processes. *J. Stat. Mech.*, page P08010, 2006.

[PTvNE02]   I. Pagonabarraga, E. Trizac, T. P. C. van Noije, and M. H. Ernst. Randomly driven granular fluids: Collisional statistics and short scale structure. *Phys. Rev. E*, 65(1):011303, 2002.

[PVB+05]   A. Puglisi, P. Visco, A. Barrat, E. Trizac, and F. van Wijland. Fluctuations of internal energy flow in a vibrated granular gas. *Phys. Rev. Lett.*, 95:110202, 2005.

[PVTvW06]   A. Puglisi, P. Visco, E. Trizac, and F. van Wijland. Dynamics of a tracer granular particle as a non-equilibrium markov process. *Phys. Rev. E*, 73:021301, 2006.

[Vis06]   P. Visco. Work fluctuations for a Brownian particle between two thermostats. *J. Stat. Mech.*, page P06006, 2006.

[vNE98]   T. P. C. van Noije and M. H. Ernst. Velocity distributions in homogeneously cooling and heated granular fluids. *Granular Matter*, 1(2):57–64, 1998.

[vNETP99]   T. P. C. van Noije, M. H. Ernst, E. Trizac, and I. Pagonabarraga. Randomly driven granular fluids: Large scale structure. *Phys. Rev. E*, 59:4326–4341, 1999.

[VPB+05]   P. Visco, A. Puglisi, A. Barrat, E. Trizac, and F. van Wijland. Injected power and entropy flow in a heated granular gas. *Europhys. Lett.*, 72:55–61, 2005.

[VPB+06a]   P. Visco, A. Puglisi, A. Barrat, E. Trizac, and F. van Wijland. Fluctuations of power injection in randomly driven granular gases. *J. Stat. Phys.*, 125:529–564, 2006.

[VPB+06b]   P. Visco, A. Puglisi, A. Barrat, F. van Wijland, and E. Trizac. Energy fluctuations in vibrated and driven granular gases. *Eur. Phys. J. B*, 51:37–387, 2006.

[vZC03]   R. van Zon and E. G. D. Cohen. Extension of the fluctuation theorem. *Phys. Rev. Lett.*, 91:110601, 2003.

[WM96]   D. R. M. Williams and F. C. MacKintosh. Driven granular media in one dimension: correlations and equations of state. *Phys. Rev. E*, 54(1):R9–R12, 1996.

An Extended Continuum Theory
for Granular Media

Pasquale Giovine

Summary. In a dilatant granular material with rotating grains the kinetic energy in addition to the usual translational one consists of three terms owing to the microstructural motion; in particular, it includes the rotation of granules and the dilatational expansion and contraction of the individual (compressible) grains and of the grains relative to one another. Therefore the balance and constitutive equations of the medium are obtained by considering it as a continuum with a constrained affine microstructure. Moreover, the balance of granular energy is demonstrated to be a direct consequence of the balance of micromomentum, while the dilatational and the rotational microstresses are turned out to be of different physical nature. Finally, a kinetic energy theorem implies that, locally, the power of all inertial forces is the opposite of the time-rate of change of kinetic energy plus the divergence of a flux through the boundary. The peculiar case of a suspension of rotating rigid granules puts in evidence the possibility for granular materials of supporting shear stresses through the generation of microrotational gradients.

1 Introduction

In this study we extend the continuum theory of dilatant granular materials, as developed in [31], by the consideration of possible rotations of compressible granules (see also, [1] and [40]); that theory generalized the models of perfect fluids with microstructure of Capriz in Sect. 12 of [7] and of distributed bodies of Goodman and Cowin [33].

The theory of distributed continuum proposed in [33] was widely used to study the slow flows of granular materials and, in particular, the propagation of all sort of waves, the basic equations of the equilibrium theory obtained from variational principles, the multiphase granular mixtures, the shearing flows, etc. (see, e.g., [34], [38], [45]–[48] and [22]). The material was assumed to consist of dry cohesionless compressible spheres of uniform size and the flow behaviour has required a combination of suggestions from both fluid and solid mechanics owing to the fact that the material has an essentially fluid-like behaviour, but it can also be heaped and, moreover, its bulk compressibility depends on the initial voids distribution in the reference placement (see the experimental results in [41] and [2]). An additional equation of balance for the microinertia was needed for the new independent kinematical variable introduced in [33], the volume fraction of the grains which describes the local arrangements of the grains themselves: hence granular materials are a special case of continua with microstructure [7].

In [50] and [16] it was observed that the constitutive hypotheses made in [33] raised some uncertainties: these was partially rectified in Sect. 3 of [50] and in [20] and [28], at least in the case of incompressible grains. Instead, the compressible case was extensively analysed in [30], [27] and [31]. In particular, in [30] the dynamic equations of motion was obtained, in the conservative case, from a Hamiltonian variational principle of local type for a perfect fluid with microstructure, in accordance with the fluid-like behaviour of granular materials (the preference for a Eulerian variational principle, rather than Lagrangian, was not in contrast with the previous appeal to a reference placement because the difference between the former and the latter formulation is not so peremptory for such materials (see also, [3])). The choice of the expression of the total kinetic energy and of the independent constitutive variables was made in accordance with [4] (*"... the dilatational motion consists of expansion and contraction of the individual (compressible) grains ... and of the grains relative to one another ... "*) and [21] (*"... the gradient of solid's volume fraction is not, by itself, the appropriate second geometric measure of local structure ... "*), respectively.

An interesting application of the theory in [30] was investigated in [32] for the study of seismic waves propagating through a sediment filled basin in the case of rigid grains; one of the advantages of the model, with respect to purely propagative models, was the reproduction of a nonlinear effect experimentally observed for real seismic waves: site amplification decreases as the amplitude of the incident wave increases.

In this chapter we consider a suspension of elastic spheres in a compressible gas of negligible mass; we assume a volume concentration close to that of packed particles, so that the mean free path of the particles is very short in comparison to the size of the particles themselves (as it is the case of cohesionless soil or sand with rough surface grains).

In Sects. 2 and 3 we present the model for a dilatant granular material with rotating grains and make a proposal for the kinetic energy instead to mention momentum and inertia, because it appears easier to conceive an appropriate expression of the former quantity rather than of the latter. The total kinetic energy consists of four terms: in addition to the usual translational one, there are three types of microstructural motion that are modelled by our theory, the two dilatational motions previously mentioned and the rotation of granules.

In Sects. 4 and 5 we introduce the balance equations and the principal fields for continua with affine microstructure and analyse the meaning of objectivity for change in observer for these fields.

In Sect. 6 we study the kinematical constraint of spherical microstructure by imposing that the changes in the affine microstructure are conformal and then obtain the pure field equations that rule the time evolution of the macro- and micro-motion and of the temperature; moreover, we get an equation for reactions to the constraint.

In Sect. 7 we assume the validity of a kinetic energy theorem which implies that, locally, the power of all inertial forces be the opposite of the given time-rate of change of the kinetic energy plus the divergence of the flux through the boundary. Furthermore, we define the granular temperature in our theory and recover the balance of granular energy as a direct consequence of the balance of micromomentum.

In Sect. 8 we impose constitutive postulates for a thermoelastic granular medium, deduce that the Helmholtz free energy represents a sort of potential for stresses and microstresses, and compare the results with previous theories by using comments and remarks. In particular, we observe the different physical nature of the dilatational microstress with respect to the rotational one, the former expressing a sort of internal non-local action rather than the usual connection with boundary microtractions of the latter [12].

Finally, in Sect. 9 we consider the peculiar case of a suspension of rotating rigid granules in a fluid matrix and notice that the microstructure behaves as that of a microrigid Cosserat's continua. By considering possible rotations of grains during the motion, we also show that, even when the volume fraction of the grain distribution is constant, the model predicts the possibility of supporting shear stress through the generation of microrotational gradients.

2 A First Model

The continuum model for dilatant granular materials here considered is directly referred to the models proposed in [6] and [4]. The material elements of the body are a sort of quasi-particle, that will be called a 'chunk' of material, and are thought of as envelopes which fill the body without voids between them (see also, [13]): each one consists of a grain and its immediate neighbours as it is the case of a suspension of elastic particles in a compressible fluid, whose density is considered to be negligible compared with the proper density ρ_m of the suspended particles; so the chunk mass density ρ of the body equals ρ_m times the volume fraction ν of the grains

$$\rho = \rho_m \nu, \tag{1}$$

with $\nu \in [0, 1)$.

In [31] the motions allowed within the chunk were merely expansions (or contractions) of the inclusions and radial motions of the spherical crust due to the displacements of the grain relative to the centre of mass of the element itself; neither diffusion of the grains through the envelope, nor effects of relative rotations of the elements or of the granules themselves were considered: assumptions rather limiting for this type of media, but necessary to obtain suggestions for the choice of an appropriate expression of the density per unit mass of the additional kinetic coenergy $\chi\left(\rho_m, \dot{\rho}_m\right)$ due to the microstructure and related to the kinetic energy κ_d by the Legendre transformation

$$\frac{\partial \chi}{\partial \dot{\rho}_m} \cdot \dot{\rho}_m - \chi = \kappa_d \tag{2}$$

(see [10]). In (2) the dot denotes material time derivative, i.e.,

$$\dot{\rho}_m := \frac{\partial \rho_m}{\partial \tau} + \mathbf{v} \cdot \operatorname{grad} \rho_m. \tag{3}$$

In particular, if $\mathbf{v}$ denotes the velocity of the mass centre of the element, whose local position vector is $\mathbf{x}$ at the time τ ($\mathbf{x}_*$ being the reference one), thus the total kinetic coenergy of the material in [31] is homogeneous of second degree in the macro- and micro-velocities and so equal to the total kinetic energy κ_{tot} (see again, [10]); precisely, it is:

$$\kappa_{tot} = \kappa_t + \kappa_f + \kappa_d, \tag{4}$$

with

$$\kappa_t := \frac{1}{2}\mathbf{v} \cdot \mathbf{v}, \quad \kappa_f := \frac{1}{2}\gamma(\rho)\dot{\rho}^2, \quad \kappa_d := \frac{1}{2}\alpha(\rho_m)\dot{\rho}_m^2. \tag{5}$$

In (4) κ_t is the usual translational kinetic energy related to the velocity of the centre of mass of the macro-element; κ_f is the 'fluctuation' kinetic energy associated to the 'dilatancy', as defined by Reynolds [49], by means of the motion of individual grains relative to the centre of mass, i.e., the kinetic energy due to the variations of the volume of chunk interstitial voids and expressed in terms of the rate of change of the chunk mass density ρ, with $\gamma(\rho)$ a scalar constitutive coefficient; κ_d is the 'dilatational' kinetic energy related to local expansions (or contractions) of the inclusions in the chunk and written in terms of the rate of change of the proper mass density of the grains ρ_m, with $\alpha(\rho_m)$ another scalar constitutive function.

Explicit evaluations for the constitutive functions $\gamma(\rho)$ and $\alpha(\rho_m)$ can be obtained if one imagines simple microstructural motions and peculiar geometrical shapes for chunks and/or granules. In particular, if the grains and the chunks expand or contract homogeneously with independent motions and if the envelope of the chunk is imagined as a spherical surface of radius ς containing some spherical inclusions, the grains, of radius φ, which have the same radius ς_* and φ_*, respectively, in a reference placement $\mathcal{B}_*$ of the material, we calculate the following expressions (see the Appendix):

$$\gamma(\rho) = \gamma_* \rho^{-\frac{8}{3}}, \quad \alpha(\rho_m) = \alpha_* \rho_m^{-\frac{8}{3}}, \tag{6}$$

with

$$\gamma_* = \frac{16}{351}\rho_*^{\frac{2}{3}}\varsigma_*^2, \quad \alpha_* = \frac{1}{15}\rho_{m*}^{\frac{2}{3}}\varphi_*^2 \tag{7}$$

and where, now and in the course, the subscript $(\cdot)_*$ refers to the value of the quantity in the reference placement $\mathcal{B}_*$.

When other geometric configurations of the grains and of the elements are considered, it is possible to compute more general expressions for γ and α (see Sect. 2 of [4]).

3 Rotations

Hereafter we denote by Lin^+, Sym^+ and Orth^+ the collection of second-order tensors with positive determinant, symmetric and positive definite, and proper orthogonal, respectively. Moreover, sym $\mathbf{A}$ and skw $\mathbf{A}$ are the symmetric and skew parts of a second-order tensor $\mathbf{A}$, respectively, while the spherical and deviatoric parts of $\mathbf{A}$ are defined to be, respectively,

$$\mathrm{sph}\,\mathbf{A} := \frac{1}{3}(\mathrm{tr}\,\mathbf{A})\mathbf{I} \quad \text{and} \quad \mathrm{dev}\,\mathbf{A} := \mathrm{sym}\,\mathbf{A} - \mathrm{sph}\,\mathbf{A}, \tag{8}$$

where $\mathrm{tr}\,\mathbf{A} := \mathbf{A}\cdot\mathbf{I}$ is the trace of $\mathbf{A}$ and $\mathbf{I} := (\delta_{ik})$ the identity tensor with δ_{ik} the delta of Kronecker. Also, Skw is the collection of all skew second-order tensors and Sym that of all symmetric second-order tensors, direct sum of Sph and Dev, the subspaces of spherical and traceless elements of Sym, respectively.

Now we generalize the expression of the density of dilatational kinetic energy κ_d obtained in [31], and defined in $(5)_3$, in order to allow effects of relative rotations of the compressible granules.

We suppose that each grain of the continuum is capable of an affine deformation distinct from (and independent of) the local affine deformation ensuing from the macromotion (and so not adequately modelled by the classical gradient of deformation $\mathbf{F} = \frac{\partial \mathbf{x}}{\partial \mathbf{x}_*}(\mathbf{x}_*, \tau) \in \mathrm{Lin}^+$). In particular, we assume that the microstructure of the dilatant granular material is spherical, as defined in [16], i.e., the microstructural tensor field $\mathbf{G}$ of Lin^+, describing the changes in the affine structure, is conformal:

$$\mathbf{G}(\mathbf{x}_*, \tau) = \beta(\mathbf{x}_*, \tau)\,\mathbf{R}(\mathbf{x}_*, \tau), \tag{9}$$

with $\beta(\mathbf{x}_*, \tau) > 0$ and $\mathbf{R}(\mathbf{x}_*, \tau) \in \mathrm{Orth}^+$, and the reference microinertia tensor field $\mathbf{J}_* \in \mathrm{Sym}^+$ has spherical values:

$$\mathbf{J}_*(\mathbf{x}_*) = \mu_*^2(\mathbf{x}_*)\,\mathbf{I}, \quad \forall \mathbf{x}_* \in \mathcal{B}_*. \tag{10}$$

Remark: We observe that the reference microinertia tensor $\mathbf{J}_*$ is directly related to the Euler's microinertia tensor per unit mass $\mathbf{J}$ of the generic chunk with respect to its centre of mass $\mathbf{x}$ at time τ and to the corresponding kinetic energy density κ_s, two fields which have the following form, respectively:

$$\mathbf{J} = \mathbf{G}\mathbf{J}_*\mathbf{G}^T \in \mathrm{Sym}^+ \quad \text{and} \quad \kappa_s = \frac{1}{2}(\mathbf{V}\mathbf{J}_*) \cdot \mathbf{V}, \tag{11}$$

where $\mathbf{V}(\mathbf{x}_*,\tau) := \dot{\mathbf{G}}(\mathbf{x}_*,\tau)$ is the microvelocity over the current placement $\mathcal{B}_\tau = \mathbf{x}(\mathcal{B}_*,\tau)$ of the body $\mathcal{B}$ (see e.g., (2.10) and (2.35) of [16] and, more in general, (5) and (16) of [9]).

For dilatant granular materials with rotating grains, the microstructure is supposed spherical and relations (9) and (10) apply, hence the Euler's tensor $\mathbf{J}$ is always spherical and the inertia related to the admissible micromotions of grains is decomposed in two terms because the trace of the skew tensor product $\dot{\mathbf{R}}\mathbf{R}^T$ vanishes; they are expressed by

$$\mathbf{J} = \mu^2 \mathbf{I} \quad \text{and} \quad \kappa_s = \frac{3}{2}\dot{\mu}^2 + \frac{1}{2}\mu^2 \dot{\mathbf{R}} \cdot \dot{\mathbf{R}}, \tag{12}$$

respectively, with

$$\mu(\mathbf{x}_*,\tau) := \mu_*(\mathbf{x}_*)\beta(\mathbf{x}_*,\tau). \tag{13}$$

An explicit suggestion for the constitutive expression of μ is obtained by considering the previous model of Sect. 2 as a particular case of this one; thus, by restricting the rotation $\mathbf{R}$ to coincides with the identity tensor $\mathbf{I}$, the kinetic energy κ_s must reduce to the kinetic energy κ_d of $(5)_3$ with $\alpha(\rho_m)$ given by $(6)_2$. Thus the following relation is valid by identification (in the case $\mathbf{R} = \mathbf{I}$):

$$\frac{3}{2}\dot{\mu}^2 = \frac{1}{2}\alpha_* \, \rho_m^{-\frac{8}{3}} \, \dot{\rho}_m^2; \tag{14}$$

so that a straightforward integration of the latter equation yields the following requested constitutive term:

$$\mu(\rho_m) = \mu_* + \sqrt{3\alpha_*}(\rho_m^{-\frac{1}{3}} - \rho_{m*}^{-\frac{1}{3}}). \tag{15}$$

Therefore, by choosing $\mu_* = \rho_{m*}^{-\frac{1}{3}}\sqrt{3\alpha_*}$, we have that

$$\mu(\rho_m) = \rho_m^{-\frac{1}{3}}\sqrt{3\alpha_*} \quad \text{and} \quad \beta(\rho_m) = \left(\frac{\rho_{m*}}{\rho_m}\right)^{\frac{1}{3}}, \tag{16}$$

so the conformal coefficient β accounts for the homogeneous expansion or contraction of the grains.

At the end, in this chapter the total kinetic energy κ_{ext} of the extended model for dilatant granular media is:

$$\kappa_{ext} = \frac{1}{2}\mathbf{v} \cdot \mathbf{v} + \frac{1}{2}\gamma(\rho)\dot{\rho}^2 + \frac{3}{2}\dot{\mu}^2(\rho_m) + \frac{1}{2}\mu^2(\rho_m)\dot{\mathbf{R}} \cdot \dot{\mathbf{R}}, \tag{17}$$

with $\gamma(\rho)$ and $\mu(\rho_m)$ given by $(6)_1$ and $(16)_1$, respectively.

4 Balance of Interactions for Material Bodies with Affine Microstructure

The local statements of the balance laws for granular materials will be obtained in Sect. 6 by the general ones for bodies with affine microstructure by imposing the internal constraint (9) on the tensor field $\mathbf{G}$ describing the changes in the affine structure. These equations of balance governing an admissible thermomechanical process are (see e.g., Sect. 21 of [7] and Sect. II.C of [43]):

$$\dot{\rho} + \rho \operatorname{tr} \mathbf{L} = 0, \tag{18}$$

$$\mathbf{c} + \operatorname{div} \mathbf{T} = \mathbf{0}, \tag{19}$$

$$\mathbf{C} - \mathbf{Z} + \operatorname{div} \mathbf{\Sigma} = \mathbf{0}, \tag{20}$$

$$\operatorname{skw} \mathbf{T} = \operatorname{skw} \left(\mathbf{G} \mathbf{Z}^T + \operatorname{grad} \mathbf{G} \odot \mathbf{\Sigma} \right), \tag{21}$$

$$\rho \dot{\epsilon} = \mathbf{T} \cdot \mathbf{L} + \mathbf{Z} \cdot \mathbf{V} + \mathbf{\Sigma} \cdot \operatorname{grad} \mathbf{V} + \rho \lambda - \operatorname{div} \mathbf{q}. \tag{22}$$

Equation (18) is the conservation law of mass and $\mathbf{L}$ is the usual velocity gradient: $\mathbf{L} := \operatorname{grad} \mathbf{v}\,(= \dot{\mathbf{F}} \mathbf{F}^{-1})$; equation (19) is the standard law of Cauchy's balance, where $\mathbf{c}$ is the vector density per unit volume of external bulk forces and $\mathbf{T}$ the stress tensor; equation (20) is the balance of microstructural interactions, in which $\mathbf{C}$ and $-\mathbf{Z}$ are the resultant tensor densities per unit volume of external bulk interactions on the microstructure and internal self-force, respectively, while $\mathbf{\Sigma}$ is the third-order microstress tensor that, in general, is not necessarily related to a sort of boundary microtractions, unless it is possible to define a physically significant connection on the manifold of values of the microstructure by which the gradient on it may be evaluated in covariant manner (see [11]); equation (21) is the balance law of angular momentum and the tensor product $\odot$ between third-order tensors is so defined:

$$(\operatorname{grad} \mathbf{G} \odot \mathbf{\Sigma})_{ij} := \mathbf{G}_{ih,k} \mathbf{\Sigma}_{jhk}; \tag{23}$$

equation (22) is the balance of mechanical energy in which ϵ is the specific internal energy per unit mass, λ the scalar rate of heat generation per unit mass due to irradiation and $\mathbf{q}$ the heat flux vector.

We accept here the principle of entropy as it applies in its classical form purely thermal: intrinsic production of entropy is always non-negative during every admissible thermodynamic process for the body. This production is given by the rate of variation of the specific entropy, whose density per unit mass is η, less the rate of heat exchange due to a flux of entropy through the boundary of vector density $-\theta^{-1}\mathbf{q}$, where θ is the (positive) absolute temperature, and a production owing to distributed entropy sources of specific density per unit mass $\lambda \theta^{-1}$. The local form of the principle is given by the Clausius–Duhem inequality

$$\rho \dot{\eta} + \operatorname{div} \left(\theta^{-1} \mathbf{q} \right) - \rho \lambda \theta^{-1} \geq 0; \tag{24}$$

moreover, if we introduce the Helmholtz free energy per unit mass $\psi := \epsilon - \theta\eta$ and use (22), we obtain a reduced version of this inequality, that is,

$$\rho\left(\dot{\psi} + \dot{\theta}\eta\right) + \theta^{-1}\mathbf{q} \cdot \mathbf{g} \leq \mathbf{T} \cdot \mathbf{L} + \mathbf{Z} \cdot \mathbf{V} + \boldsymbol{\Sigma} \cdot \operatorname{grad}\mathbf{V}. \tag{25}$$

where $\mathbf{g} := \operatorname{grad}\theta$.

Equations (20) and (21) are not immediately recognized to be the balance equations which are usually proposed for studying continua with affine microstructure (see e.g., [17]) or micromorphic media (see e.g., [26]), but, *modulo* some innocuous changes in notation and, after, by considering the effects of inertia of possible internal vibrations of the substructures, we can recover them.

Firstly, by transposing the balance equation of micromomentum (20) and multiplying both sides by the microstructural tensor variable $\mathbf{G}$, we have the following result:

$$\mathbf{G}\mathbf{C}^T - \mathbf{G}\mathbf{Z}^T - \operatorname{grad}\mathbf{G} \odot \boldsymbol{\Sigma} + \operatorname{div}\left(\mathbf{G} \oslash {}^t\boldsymbol{\Sigma}\right) = \mathbf{0}, \tag{26}$$

where the minor left transposition (of exponent t) on a tensor $\boldsymbol{\Omega}$ of the third order has the following meaning: $(({}^t\boldsymbol{\Omega}\,\mathbf{a})\mathbf{b})\mathbf{c} = ((\boldsymbol{\Omega}\,\mathbf{a})\mathbf{c})\mathbf{b}$, for each triple of vectors $\mathbf{a}$, $\mathbf{b}$ and $\mathbf{c}$, while the tensor product $\oslash$ between tensors of the second and the third order is so defined: $(\mathbf{A} \oslash \boldsymbol{\Omega})_{ijl} := \mathbf{A}_{ih}\boldsymbol{\Omega}_{hjl}$.

Then, by using the balance equation of moment of momentum (21) and by introducing the following second- and third-order tensors

$$\widetilde{\mathbf{C}} := \mathbf{G}\mathbf{C}^T, \quad \widetilde{\mathbf{Z}} := \operatorname{sym}\left[\mathbf{G}\mathbf{Z}^T + \operatorname{grad}\mathbf{G} \odot \boldsymbol{\Sigma}\right] \text{ and } \widetilde{\boldsymbol{\Sigma}} := \mathbf{G} \oslash {}^t\boldsymbol{\Sigma} \tag{27}$$

into (26), it becomes

$$\widetilde{\mathbf{C}} - \widetilde{\mathbf{Z}} - \operatorname{skw}\mathbf{T} + \operatorname{div}\widetilde{\boldsymbol{\Sigma}} = \mathbf{0}. \tag{28}$$

Secondly, we decompose the volume forces $\mathbf{C}$ in their inertial $\mathbf{C}^{in}$ and noninertial $\rho\mathbf{B}$ parts as

$$\mathbf{C} = \mathbf{C}^{in} + \rho\mathbf{B} \tag{29}$$

and observe that $\mathbf{C}^{in}$ is the opposite of the Lagrangian derivative of the microstructural kinetic coenergy $\chi_s(\mathbf{V})$, homogeneous of second degree in the micro-velocity $\mathbf{V}$ and so equal to the kinetic energy κ_s defined in $(11)_2$, thus it is

$$\mathbf{C}^{in} = -\rho\left[\frac{d}{d\tau}\left(\frac{\partial\kappa_s}{\partial\mathbf{V}}\right) - \frac{\partial\kappa_s}{\partial\mathbf{G}}\right] = -\rho\dot{\mathbf{V}}\mathbf{J}_* \tag{30}$$

(see also, [10] or (60) of [43]).

Hence, by transposing this relation, multiplying both sides by the microstructural tensor variable $\mathbf{G}$ and using relation $(11)_1$, we have that:

$$\mathbf{G}(\mathbf{C}^{in})^T = -\rho\mathbf{G}\mathbf{J}_*\dot{\mathbf{V}}^T = -\rho\mathbf{J}(\dot{\mathbf{V}}\mathbf{G}^{-1})^T, \tag{31}$$

Then, by introducing the second-order tensors $\widetilde{\mathbf{B}} := \mathbf{G}\mathbf{B}^T$ and using the relation (31) together with the (29) into (28), it becomes

$$\rho\mathbf{J}(\dot{\mathbf{V}}\mathbf{G}^{-1})^T = \rho\widetilde{\mathbf{B}} - \widetilde{\mathbf{Z}} - \text{skw}\,\mathbf{T} + \text{div}\,\widetilde{\boldsymbol{\Sigma}}. \tag{32}$$

Finally, let us insert the second-order kinematical tensor $\mathbf{W}$ for the micromotion corresponding to the velocity gradient $\mathbf{L}$ of the macromotion, i.e., the *wrenching* tensor

$$\mathbf{W}(\mathbf{x}_*, \tau) := \mathbf{V}(\mathbf{x}_*, \tau)\mathbf{G}^{-1}(\mathbf{x}_*, \tau); \tag{33}$$

for relation $(11)_1$ it satisfies the kinematical relation

$$\dot{\mathbf{J}} = \mathbf{J}\mathbf{W}^T + \mathbf{W}\mathbf{J} \tag{34}$$

that some Author calls the new fundamental conservation equation of microinertia, similar, in some sense, to the continuity equation (18) for macromotion (see, e.g., Theorem 5 in [26]): here, however, it is a simple consequence of definition $(11)_1$.

By using the wrenching (33) and relation (34) into (32), we are led to the requested classical form of equation for micromomentum (4.18) of [17]:

$$\rho\left[\overline{(\mathbf{J}\mathbf{W}^T)} - \mathbf{W}\mathbf{J}\mathbf{W}^T\right] = \rho\widetilde{\mathbf{B}} - \widetilde{\mathbf{Z}} - \text{skw}\,\mathbf{T} + \text{div}\,\widetilde{\boldsymbol{\Sigma}}, \tag{35}$$

where $\widetilde{\mathbf{B}}$ is the generalized body moment, $\widetilde{\boldsymbol{\Sigma}}$ is the hyperstress and $\widetilde{\mathbf{Z}}$ represents the symmetric part of the generalized moment of interaction of the microstructure and the gross motion. By replacing these fields in (22), we obtain the related energy equation in presence of affine microstructure in the usual form (see also (2.5) of [16]):

$$\rho\dot{e} = \mathbf{T}\cdot\mathbf{L} + (\widetilde{\mathbf{Z}} + \text{skw}\,\mathbf{T})\cdot\mathbf{W}^T + \widetilde{\boldsymbol{\Sigma}}\cdot\text{grad}\left(\mathbf{W}^T\right) + \rho\lambda - \text{div}\,\mathbf{q}. \tag{36}$$

5 Observers

Now, in order to give a suitable definition of a continuum with microstructure subject to internal kinematical constraints, as (9) and (10) are, and to study the consequences of them on the balance equations (18)–(22) and (25), we need an objective version of the total power density of mechanical internal actions $\mathcal{P}_{int}$ acting on the body $\mathcal{B}$, that is the quantity appearing, with the opposite sign, in the right-hand side of the reduced version of the imbalance of entropy (25) (see Sect. 3 of [31]):

$$\mathcal{P}_{int} = -\left(\mathbf{T}\cdot\mathbf{L} + \mathbf{Z}\cdot\mathbf{V} + \boldsymbol{\Sigma}\cdot\text{grad}\,\mathbf{V}\right). \tag{37}$$

A change in observer of a body with affine microstructure $\mathcal{B}$ relates two processes $(\check{\mathbf{x}}, \check{\mathbf{G}}, \check{\theta})(\tau)$ and $(\mathbf{x}, \mathbf{G}, \theta)(\tau)$ if, for any $(\mathbf{x}_*, \tau) \in \mathcal{B}_* \times \Re$,

$$\check{\mathbf{x}}(\mathbf{x}_*, \tau) = \mathbf{c}(\tau) + \mathbf{Q}(\tau)\,\mathbf{x}(\mathbf{x}_*, \tau), \quad \check{\mathbf{G}}(\mathbf{x}_*, \tau) = \mathbf{Q}(\tau)\,\mathbf{G}(\mathbf{x}_*, \tau) \qquad (38)$$

and

$$\check{\theta}(\mathbf{x}_*, \tau) = \theta(\mathbf{x}_*, \tau), \qquad (39)$$

where $\mathbf{c}$ is a vector and $\mathbf{Q}$ a proper orthogonal tensor of Orth^+.

This means that $\mathbf{G}$ transforms like the deformation gradient $\mathbf{F}$ and can be considered as a double vector, while the velocity $\mathbf{v}$, the microvelocity $\mathbf{V}$ and the gradient of temperature $\mathbf{g}$ transform as follows:

$$\check{\mathbf{v}} = \dot{\mathbf{c}} + \dot{\mathbf{Q}}\mathbf{d} + \mathbf{Q}\mathbf{v}, \quad \check{\mathbf{V}} = \dot{\mathbf{Q}}\mathbf{G} + \mathbf{Q}\mathbf{V} \quad \text{and} \quad \check{\mathbf{g}} = \mathbf{Q}\mathbf{g}, \qquad (40)$$

where $\mathbf{d}$ is the position vector of $\mathbf{x}$ relative to a fixed origin in $\mathcal{E}$.

Now, let $\mathbf{D}\,(:= \mathrm{sym}\,\mathbf{L})$ and $\mathbf{Y}\,(:= -\mathrm{skw}\,\mathbf{L})$ be the stretching and the spin tensor, respectively, and $\tilde{\mathbf{D}}\,(:= \mathrm{sym}\,\mathbf{W})$ and $\tilde{\mathbf{Y}}\,(:= -\mathrm{skw}\,\mathbf{W})$ the micro-stretching and the micro-spin tensor, respectively, so that

$$\mathbf{L} = \mathbf{D} - \mathbf{Y} \quad \text{and} \quad \mathbf{W} = \tilde{\mathbf{D}} - \tilde{\mathbf{Y}}; \qquad (41)$$

therefore one can compute from $(40)_{1,2}$ the transformation laws for $\check{\mathbf{L}}$ and $\check{\mathbf{W}}$:

$$\check{\mathbf{L}} = \overline{\mathrm{grad}\,\check{\mathbf{v}}} = \frac{\partial \check{\mathbf{L}}}{\partial \mathbf{x}_*}\check{\mathbf{F}}^{-1} = \left(\dot{\mathbf{Q}}\mathbf{F} + \mathbf{Q}\frac{\partial \mathbf{v}}{\partial \mathbf{x}_*}\right)\mathbf{F}^{-1}\mathbf{Q}^T = \dot{\mathbf{Q}}\mathbf{Q}^T + \mathbf{Q}\mathbf{L}\mathbf{Q}^T \quad (42)$$

and

$$\check{\mathbf{W}} = \check{\mathbf{V}}\check{\mathbf{G}}^{-1} = \left(\dot{\mathbf{Q}}\mathbf{G} + \mathbf{Q}\mathbf{V}\right)\mathbf{G}^{-1}\mathbf{Q}^T = \dot{\mathbf{Q}}\mathbf{Q}^T + \mathbf{Q}\mathbf{W}\mathbf{Q}^T; \qquad (43)$$

consequently, $\check{\mathbf{L}}$ can be split into the symmetric and skew part, respectively:

$$\check{\mathbf{D}} = \mathbf{Q}\mathbf{D}\mathbf{Q}^T \quad \text{and} \quad \check{\mathbf{Y}} = -\dot{\mathbf{Q}}\mathbf{Q}^T - \mathbf{Q}\mathbf{Y}\mathbf{Q}^T, \qquad (44)$$

as well as $\check{\mathbf{W}}$.

Owing to the transformation laws $(40)_2$ and $(42)_4$, the expression (37) for the power density $\mathcal{P}_{int}$ is not frame indifferent, apparently; instead, by using the balance of angular momentum (21) and relations (41), we have that

$$\begin{aligned}
-\mathcal{P}_{int} &= \mathbf{D} \cdot \mathrm{sym}\,\mathbf{T} + \mathbf{Y} \cdot \mathrm{skw}\left(\mathbf{Z}\mathbf{G}^T + \mathbf{\Sigma} \odot \mathrm{grad}\,\mathbf{G}\right) + \\
&\quad + \mathbf{W} \cdot \left(\mathbf{Z}\mathbf{G}^T + \mathbf{\Sigma} \odot \mathrm{grad}\,\mathbf{G}\right) + (\mathbf{G} \oslash {}^t\mathbf{\Sigma}) \cdot \mathrm{grad}\,(\mathbf{W}^T) = \\
&= \mathbf{D} \cdot \mathrm{sym}\,\mathbf{T} + (\mathbf{Y} - \tilde{\mathbf{Y}}) \cdot \mathrm{skw}\left(\mathbf{Z}\mathbf{G}^T + \mathbf{\Sigma} \odot \mathrm{grad}\,\mathbf{G}\right) + \\
&\quad + \tilde{\mathbf{D}} \cdot \mathrm{sym}\left(\mathbf{Z}\mathbf{G}^T + \mathbf{\Sigma} \odot \mathrm{grad}\,\mathbf{G}\right) + (\mathbf{G} \oslash {}^t\mathbf{\Sigma}) \cdot \mathrm{grad}\,(\tilde{\mathbf{D}} + \tilde{\mathbf{Y}})
\end{aligned} \qquad (45)$$

and hence $\mathcal{P}_{int}$ is indifferent to changes in observer, as requested.

6 Dilatant Granular Materials with Rotating Grains

We now impose the perfect kinematical constraint of spherical microstructure, as described by formulas (9) and (10), in order to obtain the balance laws for granular materials which allow effects of microrotation of the compressible grains, other than the dilatancy of the chunks.

The body $\mathcal{B}$ is said to be *internally constrained* if the allowed velocity, microvelocity and temperature gradient distributions are such that not all values of the objective factors $\mathbf{D}$, $\widetilde{\mathbf{D}}$, $(\mathbf{Y} - \widetilde{\mathbf{Y}})$, $\mathrm{grad}\,\widetilde{\mathbf{D}}$, $\mathrm{grad}\,\widetilde{\mathbf{Y}}$ and $\mathbf{g}$ are accessible. In our case the wrenching $\mathbf{W}$, the micro-stretching $\widetilde{\mathbf{D}}$ and the micro-spin $\widetilde{\mathbf{Y}}$ are given by

$$\mathbf{W} = \dot{\beta}\beta^{-1}\mathbf{I} + \dot{\mathbf{R}}\mathbf{R}^T, \quad \widetilde{\mathbf{D}} = \dot{\beta}\beta^{-1}\mathbf{I}, \quad \widetilde{\mathbf{Y}} = -\dot{\mathbf{R}}\mathbf{R}^T, \tag{46}$$

respectively, and so the macromotion is not constrained at all, while

$$\mathrm{grad}\,(\mathbf{W}^T) = \mathbf{I} \otimes \mathrm{grad}\,(\dot{\beta}\beta^{-1}) + \mathrm{grad}\,(\dot{\mathbf{R}}\mathbf{R}^T). \tag{47}$$

Furthermore, we follow classical theories (see [36] and [18]) and suppose that each quantity, which, in absence of the constraint, is ruled by a constitutive prescription (that is $\mathbf{T}$, $\mathbf{Z}$, $\mathbf{\Sigma}$, $\mathbf{q}$, ϵ, η, ψ) is now the direct sum of one *active* and one *reactive* component

$$\mathbf{T} = \mathbf{T}_a + \mathbf{T}_r,\ \mathbf{Z} = \mathbf{Z}_a + \mathbf{Z}_r,\ \text{etc.} \tag{48}$$

and only the active component is bound through suitable constitutive relations to the independent thermokinetic variables.

The additional request that the constraint is *perfect*, i.e., internally frictionless, is specified, in this wider thermomechanic rather than purely mechanical context, by the property that the entropy production due to the reaction is null, that is the contribution of the reactions to the inequality (25) are identically zero for every process allowed by the constraint (see also, Sect. 27 of [7]):

$$\rho\left(\dot{\psi}_r + \eta_r\dot{\theta}\right) + \theta^{-1}\mathbf{q}_r \cdot \mathbf{g} = \mathbf{T}_r \cdot \mathbf{L} + \mathbf{Z}_r \cdot \mathbf{V} + \mathbf{\Sigma}_r \cdot \mathrm{grad}\,\mathbf{V}. \tag{49}$$

By using the representation $(45)_1$ of $\mathcal{P}_{int}$, the constraint relation (9), (46) and (47) into (49), we have

$$\rho\left(\dot{\psi}_r + \eta_r\dot{\theta}\right) + \theta^{-1}\mathbf{q}_r \cdot \mathbf{g} = \mathrm{sym}\,\mathbf{T}_r \cdot \mathbf{D} - \mathrm{skw}\,\mathbf{T}_r \cdot \mathbf{Y} +$$

$$+ (\dot{\beta}\beta^{-1})\,[\beta\mathbf{Z}_r \cdot \mathbf{R} + \mathbf{\Sigma}_r \cdot \mathrm{grad}\,(\beta\mathbf{R})] + \left(\beta\mathbf{\Sigma}_r^T\,\mathbf{R}^T\right) \cdot \mathrm{grad}\,\left(\dot{\beta}\beta^{-1}\right) - \tag{50}$$

$$- \mathrm{skw}\,\left[\beta\mathbf{Z}_r\mathbf{R}^T + \mathbf{\Sigma}_r \odot \mathrm{grad}\,(\beta\mathbf{R})\right] \cdot \widetilde{\mathbf{Y}} + (\beta\mathbf{R} \oslash {}^t\mathbf{\Sigma}_r) \cdot \mathrm{grad}\,\widetilde{\mathbf{Y}},$$

for every totally free choice of $\dot{\theta}$ and $(\dot{\beta}\beta^{-1})$ among the scalars, $\mathbf{g}$ among the vectors, $\mathbf{Y}$ and $\widetilde{\mathbf{Y}}$ among the skew tensors and $\mathbf{D}$ among the symmetric tensors; in (50) the transposition of exponent T on a tensor $\boldsymbol{\Omega}$ of the third order has the following meaning: $((\boldsymbol{\Omega}^T\mathbf{a})\mathbf{b})\mathbf{c} = ((\boldsymbol{\Omega}\,\mathbf{c})\mathbf{b})\mathbf{a}$, for each triple of vectors $\mathbf{a}$, $\mathbf{b}$ and $\mathbf{c}$.

The reactions are then characterized by the following requirements:

$$\psi_r = \text{const.}, \quad \eta_r = 0, \quad \mathbf{q}_r = \mathbf{0}, \tag{51}$$

$$\mathbf{T}_r = \mathbf{0}, \quad \mathbf{Z}_r \cdot \beta\mathbf{R} + \boldsymbol{\Sigma}_r \cdot \text{grad}\,(\beta\mathbf{R}) = 0,$$

$$\text{skw}\left[\beta\mathbf{Z}_r\mathbf{R}^T + \boldsymbol{\Sigma}_r \odot \text{grad}\,(\beta\mathbf{R})\right] = \mathbf{0}, \tag{52}$$

$$\boldsymbol{\Sigma}_r^T\,\mathbf{R}^T = \mathbf{0} \quad \text{and} \quad \text{skw}\left[\beta\mathbf{R}\,(\boldsymbol{\Sigma}_r\mathbf{w})^T\right] = \mathbf{0}, \ \forall\,\text{vector}\,\mathbf{w};$$

hence, from definitions $(27)_{2,3}$, we have that reactions must be such that

$$\widetilde{\mathbf{Z}}_r = \left[\beta\,\mathbf{Z}_r\mathbf{R}^T + \boldsymbol{\Sigma}_r \odot \text{grad}\,(\beta\mathbf{R})\right] \in \text{Dev},$$

$$\widetilde{\boldsymbol{\Sigma}}_r\,\mathbf{w} = \beta\,\mathbf{R}\,(\boldsymbol{\Sigma}_r\,\mathbf{w})^T \in \text{Dev}, \ \forall\,\text{vector}\,\mathbf{w}, \tag{53}$$

and, accordingly,

$$\widetilde{\mathbf{Z}}_a \in \text{Sph} \quad \text{and} \quad \left(\widetilde{\boldsymbol{\Sigma}}_a\,\mathbf{w}\right) \in \text{Sph} \oplus \text{Skw}, \ \forall\,\text{vector}\,\mathbf{w}, \tag{54}$$

while, for $(52)_1$, $\mathbf{T}_a$ is a free tensor field, not necessarily symmetric-valued.

Now we are able to obtain a set of *pure* equations which rules the thermomechanical evolution of our model of dilatant granular material $\mathcal{B}$; in fact, by splitting the stress tensor $\mathbf{T}$ into its symmetric and skew parts and by using the condition $(52)_1$ into $(48)_1$, together with the balance of moment of momentum (21) and condition $(52)_3$, the following reaction-free expression for the stress $\mathbf{T}$ follows:

$$\mathbf{T} = \text{sym}\,\mathbf{T}_a + \text{skw}\left[\beta\mathbf{R}\mathbf{Z}_a^T + \text{grad}\,(\beta\mathbf{R}) \odot \boldsymbol{\Sigma}_a\right], \tag{55}$$

which will be the object of a constitutive prescription and it is clearly not symmetric, in general.

Moreover, by using relations (53) and (54), the balance for micromomentum in the shape (28), broken up into spherical, skew and deviatoric part, delivers

$$\text{sph}\left(\widetilde{\mathbf{C}} - \widetilde{\mathbf{Z}}_a + \text{div}\,\widetilde{\boldsymbol{\Sigma}}_a\right) = \mathbf{0}, \quad \text{skw}\left(\widetilde{\mathbf{C}} - \mathbf{T} + \text{div}\,\widetilde{\boldsymbol{\Sigma}}_a\right) = \mathbf{0} \tag{56}$$

$$\text{and} \quad \widetilde{\mathbf{Z}}_r - \text{div}\,\widetilde{\boldsymbol{\Sigma}}_r = \text{dev}\,\widetilde{\mathbf{C}}, \tag{57}$$

respectively; therefore, the constraint (9), definitions (27) and the skew part of the stress tensor furnished by (55) permit us to write the following equations:

$$\beta\,(\mathbf{C} - \mathbf{Z}_a + \text{div}\,\boldsymbol{\Sigma}_a) \cdot \mathbf{R} = 0, \ \beta\,\text{skw}\left[(\mathbf{C} - \mathbf{Z}_a + \text{div}\,\boldsymbol{\Sigma}_a)\,\mathbf{R}^T\right] = \mathbf{0} \tag{58}$$

$$\text{and} \quad \mathbf{Z}_r - \text{div}\,\boldsymbol{\Sigma}_r = \left[\text{dev}\,(\mathbf{C}\,\mathbf{R}^T)\right]\mathbf{R}. \tag{59}$$

In conclusion, only the active constitutive components of the fields of stress, internal actions and microstress appear in the Cauchy equation (19), with $\mathbf{T}$ given by (55), and in the spherical and skew parts of equation for micromomentum (58): these are the pure equations which rule the mechanical evolution of the body.

Once a motion is ensued from them, the corresponding reactions to the constraint are obtained by the condition (59) (other than by (51)) within the intrinsic indeterminacy generated from equation itself for $\mathbf{Z}_r$ and $\mathbf{\Sigma}_r$, as pointed out in Sects. 205 and 227 of [51] or in Remark 1, Sect. 3 of [15].

Now let us use the definition of the Helmholtz free energy ψ and the results (51), (52) and (55) in the balance equation for energy (22); on repeating the same procedure leading to (50), we immediately get

$$\rho\overline{(\dot{\psi}_a + \theta\eta_a)} = \mathbf{D} \cdot \operatorname{sym} \mathbf{T}_a + \left(\dot{\beta}\beta^{-1}\right) \left[\beta \mathbf{Z}_a \cdot \mathbf{R} + \mathbf{\Sigma}_a \cdot \operatorname{grad}\left(\beta \mathbf{R}\right)\right] +$$

$$+ \left(\mathbf{Y} + \dot{\mathbf{R}}\mathbf{R}^T\right) \cdot \operatorname{skw} \left[\beta \mathbf{Z}_a \mathbf{R}^T + \mathbf{\Sigma}_a \odot \operatorname{grad}\left(\beta \mathbf{R}\right)\right] + \qquad (60)$$

$$+ \left(\beta \mathbf{\Sigma}_a^T \mathbf{R}^T\right) \cdot \operatorname{grad}\left(\dot{\beta}\beta^{-1}\right) - \left(\beta \mathbf{R} \oslash {}^t\mathbf{\Sigma}_a\right) \cdot \operatorname{grad}\left(\dot{\mathbf{R}}\mathbf{R}^T\right) + \rho\lambda - \operatorname{div} \mathbf{q}_a,$$

where there is no trace of effects due to the constraint: we have obtained the pure equation of evolution for the temperature of the body.

We observe that (60) will be greatly simplified when the constitutive prescriptions for the active fields will be given and the consequences of the Clausius–Duhem inequality (25) will be taken into account.

7 Inertia Forces and Balance of Granular Energy

The fundamental pure equations of balance (19) (with the stress tensor $\mathbf{T}$ given by (55)), (58) and (60) presented in the previous sections apply to the general class of materials with spherical microstructure.

The material properties of granular media are assigned through constitutive hypotheses of thermomechanic and kinematical character: the former will be rendered explicit in the next section with the choice of constitutive postulates for a thermoelastic continuum; the latter involve the delicate argument of the connection between an appropriate choice of the densities of macro- and micro-structural inertia forces and the chosen expression (17) for the total kinetic energy density κ_{ext}.

We follow Mariano [43] and Capriz [7] and, firstly, decompose the volume force density $\mathbf{c}$ in its inertial $\mathbf{c}^{in}$ and noninertial $\rho\mathbf{f}$ part, as made in (29) for $\mathbf{C}$:

$$\mathbf{c} = \mathbf{c}^{in} + \rho\mathbf{f}; \qquad (61)$$

after we assume the validity of a kinetic energy theorem, which implies that, locally, the power for unit volume of inertial forces be the opposite of the time-rate of change of the kinetic energy density per unit mass κ_{ext}, times ρ, plus the divergence of the flux of kinetic energy density $\mathbf{k}$ through the boundary, that is

$$\mathbf{c}^{in} \cdot \mathbf{v} + \mathbf{C}^{in} \cdot \mathbf{V} = -\rho\,\dot{\kappa}_{ext} + \operatorname{div}\mathbf{k}. \tag{62}$$

It is easy to check that

$$\rho\,\dot{\kappa}_{ext} = \rho\left[\mathbf{v}\cdot\dot{\mathbf{v}} + \dot{\rho}\left(\gamma(\rho)\ddot{\rho} + \tfrac{1}{2}\gamma'(\rho)\dot{\rho}^2\right) + 3\dot{\mu}\ddot{\mu} + \mu\dot{\mu}\,\dot{\mathbf{R}}\cdot\dot{\mathbf{R}} + \mu^2\,\dot{\mathbf{R}}\cdot\ddot{\mathbf{R}}\right] =$$
$$= \left\{\rho\dot{\mathbf{v}} + \operatorname{grad}\left[\rho^2\left(\gamma(\rho)\ddot{\rho} + \tfrac{1}{2}\gamma'(\rho)\dot{\rho}^2\right)\right]\right\}\cdot\mathbf{v} + \tag{63}$$
$$+\ \mu_*\rho\left(\ddot{\mu}\mathbf{R} + 2\dot{\mu}\dot{\mathbf{R}} + \mu\ddot{\mathbf{R}}\right)\cdot\left(\dot{\beta}\mathbf{R} + \beta\dot{\mathbf{R}}\right) - \operatorname{div}\left[\rho^2\left(\gamma(\rho)\ddot{\rho} + \tfrac{1}{2}\gamma'(\rho)\dot{\rho}^2\right)\mathbf{v}\right],$$

where the continuity equation (18), the relation (13) and the properties $\mathbf{R}\cdot\mathbf{R} = 3$ and $\mathbf{R}\cdot\dot{\mathbf{R}} = 0$ of the orthogonal tensor $\mathbf{R}$ are used; the prime $(\cdot)'$ denotes differentiation with respect to the argument. Therefore, it must be:

$$\mathbf{c}^{in} = -\rho\,\dot{\mathbf{v}} - \operatorname{grad}\left[\rho^2\left(\gamma(\rho)\ddot{\rho} + \tfrac{1}{2}\gamma'(\rho)\dot{\rho}^2\right)\right], \tag{64}$$
$$\mathbf{C}^{in} = -\mu_*\rho\left(\ddot{\mu}\mathbf{R} + 2\dot{\mu}\dot{\mathbf{R}} + \mu\ddot{\mathbf{R}}\right)\ \text{and}\ \mathbf{k} = -\rho^2\left(\gamma(\rho)\ddot{\rho} + \tfrac{1}{2}\gamma'(\rho)\dot{\rho}^2\right)\mathbf{v}.$$

We observe that, for the constraints (9) and (10), $\mathbf{C}^{in}$ satisfies again relation $(30)_2$, while the expression $(64)_1$ for $\mathbf{c}^{in}$ was already obtained in Appendix B of [28] and Sect. 3 of [30] with variational procedures.

Furthermore, we also note from $(64)_1$ that there is a contribution to the total Cauchy stress tensor $\widetilde{\mathbf{T}}$ in addition to the classical surface actions exerted through the boundary and coming from an influx of linear momentum described by a tensor of inertia flux $\mathbf{M}$ which is the Lagrangian derivative, times $\rho\mathbf{I}$, of the fluctuation energy κ_f and measures the agitation within a chunk of material (see [29]): hence, it is

$$\widetilde{\mathbf{T}} = \mathbf{T} + \mathbf{M} \quad\text{with}\quad \mathbf{M} := -\rho^2\left[\gamma(\rho)\ddot{\rho} + \frac{1}{2}\gamma'(\rho)\dot{\rho}^2\right]\mathbf{I} \tag{65}$$

with $\mathbf{T}$ given by (55) (see also, the collisional–translational contribution to the total stress tensor in (2.6) of [39] or, for (18), the spherical part of a type of Reynolds stress tensor of the turbulence theory in (3.14) of the review paper [37], in which many other granular theories that split the stress tensor are examined); therefore, by using (61), the Cauchy equation in this context is so written:

$$\rho\dot{\mathbf{v}} = \rho\mathbf{f} + \operatorname{div}\widetilde{\mathbf{T}}. \tag{66}$$

From relations (29), $(64)_2$ and (13) and the properties of $\mathbf{R}$, it follows that the frictionless micromomentum balances (58) are now:

$$\rho\left(3\mu\ddot{\mu} - \mu^2\widetilde{\mathbf{Y}}\cdot\widetilde{\mathbf{Y}}\right) = \beta\mathbf{R}\cdot\left(\rho\mathbf{B} - \mathbf{Z}_a + \operatorname{div}\mathbf{\Sigma}_a\right) \quad\text{and}$$
$$\rho\left(\mu^2\widetilde{\mathbf{Y}}\right) = \operatorname{skw}\left[\beta\mathbf{R}\left(\rho\mathbf{B} - \mathbf{Z}_a + \operatorname{div}\mathbf{\Sigma}_a\right)^T\right] \tag{67}$$

or, by inserting the constitutive expressions (16) and (6)$_2$,

$$3\rho\rho_m^{\frac{4}{3}}\left[\alpha(\rho_m)\left(\ddot{\rho}_m - \rho_m\widetilde{\mathbf{Y}}\cdot\widetilde{\mathbf{Y}}\right) + \tfrac{1}{2}\alpha'(\rho_m)\dot{\rho}_m^2\right] = \rho_{m*}^{\frac{1}{3}}\mathbf{R}\cdot(\rho\mathbf{B} - \mathbf{Z}_a + \operatorname{div}\mathbf{\Sigma}_a)$$

$$\text{and}\quad 3\rho\rho_m^{\frac{1}{3}}\left(\rho_m^2\alpha(\rho_m)\widetilde{\mathbf{Y}}\right)^{\cdot} = \rho_{m*}^{\frac{1}{3}}\operatorname{skw}\left[\mathbf{R}\left(\rho\mathbf{B} - \mathbf{Z}_a + \operatorname{div}\mathbf{\Sigma}_a\right)^T\right].\quad(68)$$

Equation (59) for the reactions is now

$$\mathbf{Z}_r - \operatorname{div}\mathbf{\Sigma}_r = \rho\left[\operatorname{dev}\left(\mathbf{B}\,\mathbf{R}^T + \mu_*\mu^2\widetilde{\mathbf{Y}}^2\right)\right]\mathbf{R},\quad\text{or}$$

$$\mathbf{Z}_r - \operatorname{div}\mathbf{\Sigma}_r = \rho\left[\operatorname{dev}\left(\mathbf{B}\,\mathbf{R}^T + 3\rho_{m*}^{-\frac{1}{3}}\rho_m^{\frac{7}{3}}\alpha(\rho_m)\widetilde{\mathbf{Y}}^2\right)\right]\mathbf{R}.\tag{69}$$

In the sequel of this section, we recover the relation of evolution for the granular temperature of the body (the granular heat transfer equation (4.6) of [13] or the balance of pseudo-thermal energy (2.7) of [39]) as a direct consequence of our equations for micromomentum balance (67).

The quantity that is usually introduced as granular temperature ϑ represents a fraction of the extra energy due to grains agitation and to chunks dilatancy (and is the trace of the so-called Reynolds tensor which measures the momentum flux in fluid dynamics); in our theory it corresponds to the fluctuation energy κ_f plus the roto-dilatational kinetic energy κ_s (multiplied by $\tfrac{2}{3}$):

$$\vartheta := \frac{2}{3}(\kappa_f + \kappa_s) = \frac{1}{3}\gamma(\rho)\dot{\rho}^2 + \dot{\mu}^2(\rho_m) + \frac{1}{3}\mu^2(\rho_m)\widetilde{\mathbf{Y}}\cdot\widetilde{\mathbf{Y}},\tag{70}$$

where relation (46)$_3$ was used.

By differentiating with respect to the time and by using (17), (63) and (18), the representation (65)$_2$ of the inertia flux tensor $\mathbf{M}$ and the antisymmetry of $\widetilde{\mathbf{Y}}$, we obtain

$$\frac{3}{2}\rho\dot{\vartheta} = \mathbf{M}\cdot\mathbf{L} + \rho\dot{\mu}\mu^{-1}\left(3\mu\ddot{\mu} - \mu^2\widetilde{\mathbf{Y}}\cdot\widetilde{\mathbf{Y}}\right) + \rho\widetilde{\mathbf{Y}}\cdot\left(\mu^2\widetilde{\mathbf{Y}}\right)^{\cdot};\tag{71}$$

at the end, the equations for micromomentum balance (67) give

$$\frac{3}{2}\rho\dot{\vartheta} = \operatorname{div}\mathbf{u} + \mathbf{M}\cdot\mathbf{L} + \iota + \rho\mathbf{B}\cdot(\beta\mathbf{R})^{\cdot}.\tag{72}$$

Equation (72) is the so-called balance of granular energy in which we can easily recognize, with appropriate identifications, usual terms introduced in granular theories: $\mathbf{u} := \left[\mathbf{\Sigma}_a^T\left(\beta\mathbf{R}^T\right)^{\cdot}\right]$ is the granular heat flux vector, an interstitial work flux of mechanical nature, in excess of the usual flux due to surface tractions, owing to interactions between chunks and due to grains–boundary collisions or to exchange of granules through the chunk boundary itself as well as to weakly nonlocal spatial effects (see [31], [24], [25] and [23]); $(\mathbf{M}\cdot\mathbf{L})$ is the rate of working of the inertia component of the stress tensor; $\left[\rho\mathbf{B}\cdot(\beta\mathbf{R})^{\cdot}\right]$ is a granular heat source, that some author call the 'stir' due to external actions; $\iota := -\left[\mathbf{Z}_a\cdot(\beta\mathbf{R})^{\cdot}+\mathbf{\Sigma}_a\cdot\operatorname{grad}(\beta\mathbf{R})^{\cdot}\right]$ is the local rate of dissipation due to the inelastic nature of collisions between particles, dissipation which also appears, when (9) and $(52)_{2,3,4,5}$ are taken into account, on the right-hand side of the balance of thermic internal energy (22) with the opposite sign (see also, (2.4) of [39]).

8 Constitutive Restrictions in the Thermoelastic Case

The peculiar flow behaviour of granular materials can be considered similar to fluid one, except that its bulk compressibility and temperature distribution depend on the initial porosity (see e.g., [2] and the experimental results in [4]) and thus the medium has a preferred reference placement with respect to volume distribution.

Therefore, we assume that the overall response of a thermoelastic dilatant granular materials with rotating grains depends on the set $\mathcal{S} \equiv \{\rho_*, \rho, \mathbf{s} := \operatorname{grad}\rho, \mathbf{S} := \operatorname{grad}^2\rho, \beta, \mathbf{p} := \operatorname{grad}\beta, \mathbf{P} := \operatorname{grad}^2\beta, \mathbf{R}, \mathbf{\Pi} := \operatorname{grad}\mathbf{R}, \theta_*, \theta, \mathbf{g}\}$. The symmetric tensors $\mathbf{S}$ and $\mathbf{P}$ are inserted among variables not only for consistency with the results of the conservative case in absence of rotations [30], but also because they seem the appropriate second geometric measures of local structure, namely, a sort of rough measurements of anisotropy of grains and chunks distributions, respectively (see, also, [21] and [44]).

The equipresence principle requires that each dependent constitutive field is given by a smooth function of the set $\mathcal{S}$, i.e.,

$$\{\psi_a, \eta_a, \operatorname{sym}\mathbf{T}_a, \mathbf{Z}_a, \mathbf{\Sigma}_a, \mathbf{q}_a\} = \left\{\hat{\psi}, \hat{\eta}, \hat{\mathbf{T}}, \hat{\mathbf{Z}}, \hat{\mathbf{\Sigma}}, \hat{\mathbf{q}}\right\}(\mathcal{S}); \qquad (73)$$

now let us check the compatibility of these prescriptions with the Clausius–Duhem inequality, in its reduced version (25), by incorporating the condition (49) of perfect constraint and the functional dependence of the free energy ψ_a and by using the chain rule, the conservation of mass (18) and the identities

$$\overline{\operatorname{grad}\mathbf{R}}^{\cdot} = \operatorname{grad}\dot{\mathbf{R}} - (\operatorname{grad}\mathbf{R})\mathbf{L} \quad \text{and} \quad \overline{\operatorname{grad}\omega}^{\cdot} = \operatorname{grad}\dot{\omega} - \mathbf{L}^T\operatorname{grad}\omega, \qquad (74)$$

for each scalar function ω.

We require that the entropy imbalance (25) be valid for any choice of the fields in the set $\mathcal{S}$ and their derivatives, consequently, when the terms are appropriately ordered, the inequality reads

$$\left\{\operatorname{sym}\left[\mathbf{T}_a + \rho\left(\mathbf{s}\otimes\hat{\psi}_{\mathbf{s}} + \mathbf{p}\otimes\hat{\psi}_{\mathbf{p}} + \mathbf{\Pi}^T\odot\hat{\psi}_{\mathbf{\Pi}}^T\right)\right] + \rho^2\left(\hat{\psi}_\rho + \hat{\psi}_{\mathbf{s}}\cdot\mathbf{s}\right)\mathbf{I}\right\}\cdot\mathbf{D} -$$

$$- \operatorname{skw}\left[\beta\mathbf{R}\hat{\mathbf{Z}}^T + \operatorname{grad}(\beta\mathbf{R})\odot\hat{\mathbf{\Sigma}} + \rho\left(\mathbf{s}\otimes\hat{\psi}_{\mathbf{s}} + \mathbf{p}\otimes\hat{\psi}_{\mathbf{p}} + \mathbf{\Pi}^T\odot\hat{\psi}_{\mathbf{\Pi}}^T\right)\right]\cdot\mathbf{Y} -$$

$$- \rho\left(\hat{\eta} + \hat{\psi}_\theta\right)\dot{\theta} + \left[\beta\hat{\mathbf{Z}}\cdot\mathbf{R} + \hat{\mathbf{\Sigma}}\cdot\operatorname{grad}(\beta\mathbf{R}) - \rho\beta\hat{\psi}_\beta - \rho\hat{\psi}_{\mathbf{p}}\cdot\mathbf{p}\right]\left(\dot{\beta}\beta^{-1}\right) +$$

$$+ \text{skw} \left[\beta \mathbf{R} \hat{\mathbf{Z}}^T + \text{grad}\,(\beta \mathbf{R}) \odot \hat{\mathbf{\Sigma}} + \rho \left(\hat{\psi}_{\mathbf{R}} \mathbf{R}^T + \hat{\psi}_{\mathbf{\Pi}} \odot \mathbf{\Pi} \right) \right] \cdot \tilde{\mathbf{Y}} -$$

$$- \rho \left(\hat{\psi}_{\mathbf{S}} \cdot \dot{\mathbf{S}} + \hat{\psi}_{\mathbf{P}} \cdot \dot{\mathbf{P}} + \hat{\psi}_{\mathbf{g}} \cdot \dot{\mathbf{g}} \right) + \beta \left(\hat{\mathbf{\Sigma}}^T \mathbf{R}^T - \rho \hat{\psi}_{\mathbf{p}} \right) \cdot \text{grad}\,\left(\dot{\beta} \beta^{-1} \right) +$$

$$+ \left[\mathbf{R} \oslash \left(\beta^{\,t} \hat{\mathbf{\Sigma}} - \rho^{\,t} \hat{\psi}_{\mathbf{\Pi}} \right) \right] \cdot \text{grad}\,\tilde{\mathbf{Y}} \cdot + \rho^2 \left(\mathbf{I} \otimes \hat{\psi}_{\mathbf{s}} \right) \cdot \text{grad}\,\mathbf{D} - \theta^{-1} \mathbf{q} \cdot \mathbf{g} \geq 0,$$

$$(75)$$

where subscripts denote partial differentiation with respect to the shown field, e.g., $\hat{\psi}_{\mathbf{p}} := \frac{\partial \hat{\psi}}{\partial \mathbf{p}}$.

The left-hand member of inequality (75) is linear in $\mathbf{D}$, $\mathbf{Y}$, $\dot{\theta}$, $(\dot{\beta}\beta^{-1})$, $\tilde{\mathbf{Y}}$, $\dot{\mathbf{S}}$, $\dot{\mathbf{P}}$, $\dot{\mathbf{g}}$, $\text{grad}\,\left(\dot{\beta}\beta^{-1} \right)$, $\text{grad}\,\tilde{\mathbf{Y}}$ and $\text{grad}\,\mathbf{D}$ and hence, because one can imagine, for each material element, thermomechanical processes along which these quantities take up arbitrary values at a given instant, its fulfillment implies that the coefficients in the linear expression must all vanish:

$$\hat{\psi} = \hat{\psi}(\rho_*, \rho, \beta, \mathbf{p}, \mathbf{R}, \mathbf{\Pi}, \theta_*, \theta), \quad \hat{\eta} = -\hat{\psi}_\theta,$$

$$\hat{\mathbf{T}} = -\rho \left[\rho \hat{\psi}_\rho \mathbf{I} + \text{sym}\,\left(\mathbf{p} \otimes \hat{\psi}_{\mathbf{p}} + \mathbf{\Pi}^T \odot \hat{\psi}_{\mathbf{\Pi}}^T \right) \right],$$

$$\text{skw}\,\left[\beta \mathbf{R} \hat{\mathbf{Z}}^T + \text{grad}\,(\beta \mathbf{R}) \odot \hat{\mathbf{\Sigma}} + \rho \left(\mathbf{p} \otimes \hat{\psi}_{\mathbf{p}} + \mathbf{\Pi}^T \odot \hat{\psi}_{\mathbf{\Pi}}^T \right) \right] = \mathbf{0},$$

$$\beta \hat{\mathbf{Z}} \cdot \mathbf{R} + \hat{\mathbf{\Sigma}} \cdot \text{grad}\,(\beta \mathbf{R}) = \rho \left(\beta \hat{\psi}_\beta + \hat{\psi}_{\mathbf{p}} \cdot \mathbf{p} \right), \qquad (76)$$

$$\text{skw}\,\left[\beta \mathbf{R} \hat{\mathbf{Z}}^T + \text{grad}\,(\beta \mathbf{R}) \odot \hat{\mathbf{\Sigma}} + \rho \left(\hat{\psi}_{\mathbf{R}} \mathbf{R}^T + \hat{\psi}_{\mathbf{\Pi}} \odot \mathbf{\Pi} \right) \right] = \mathbf{0},$$

$$\hat{\mathbf{\Sigma}}^T \mathbf{R}^T = \rho \hat{\psi}_{\mathbf{p}}, \quad \text{skw}\,\left\{ \mathbf{R} \left[\left(\beta \hat{\mathbf{\Sigma}} - \rho \hat{\psi}_{\mathbf{\Pi}} \right) \mathbf{w} \right]^T \right\} = \mathbf{0}, \; \forall \text{ vector } \mathbf{w},$$

while the heat flux $\hat{\mathbf{q}}$ must satisfy identically the Fourier inequality

$$\hat{\mathbf{q}} \cdot \mathbf{g} \leq 0. \qquad (77)$$

The following compatibility condition on the free energy $\hat{\psi}$, which comes out from $(76)_4$ and $(76)_6$:

$$\text{skw}\,\left(\mathbf{p} \otimes \hat{\psi}_{\mathbf{p}} + \mathbf{R} \hat{\psi}_{\mathbf{R}}^T + \mathbf{\Pi}^T \odot \hat{\psi}_{\mathbf{\Pi}}^T + \mathbf{\Pi} \odot \hat{\psi}_{\mathbf{\Pi}} \right) = \mathbf{0}, \qquad (78)$$

expresses simply the condition of frame-indifference for $\hat{\psi}$, namely,

$$\hat{\psi}\left(\rho_*, \rho, \beta, \mathbf{Qp}, \mathbf{QR}, (\mathbf{Q} \oslash \mathbf{\Pi})\,\mathbf{Q}^T, \theta_*, \theta \right) = \hat{\psi}(\rho_*, \rho, \beta, \mathbf{p}, \mathbf{R}, \mathbf{\Pi}, \dot{\theta}_*, \theta), \qquad (79)$$

for each $\mathbf{Q} \in \text{Orth}^+$.

Moreover, the total Cauchy stress tensor $\tilde{\mathbf{T}}$ for a thermoelastic medium is given by (65), (55) and $(76)_{3,4}$:

$$\widetilde{\mathbf{T}} = -\rho^2 \left[\gamma(\rho)\ddot{\rho} + \frac{1}{2}\gamma'(\rho)\dot{\rho}^2 + \hat{\psi}_\rho \right] \mathbf{I} - \rho \left(\mathbf{p} \otimes \hat{\psi}_{\mathbf{p}} + \mathbf{\Pi}^T \odot \hat{\psi}_{\mathbf{\Pi}}^T \right), \qquad (80)$$

where we recognize the usual thermodynamic pressure for fluids $\pi := \rho^2 \hat{\psi}_\rho$, related to the compressibility of granules, a stress of Ericksen's type $(-\rho \, \mathbf{p} \otimes \hat{\psi}_{\mathbf{p}})$ that justifies the ability of granular continua to support shear in equilibrium also in absence of microrotation, as evidenced by the characteristic angle of repose of these materials, and a further stress term $\left(-\rho \, \mathbf{\Pi}^T \odot \hat{\psi}_{\mathbf{\Pi}}^T \right)$, which shows that they could still sustain shear stresses when the grains are rigid, giving rise to the generation of microrotation gradients.

As observed at the end of Sect. 6, the evolution equation for the temperature of granular materials (60) simplifies considerably and reduces to the classical one, that is,

$$\rho\theta\dot{\hat{\eta}} = \rho\lambda - \operatorname{div}\hat{\mathbf{q}}. \qquad (81)$$

Furthermore, with the use of constitutive relations (76) in (73), we are able to express the dependent fields on the right-hand side of pure equations of micromotion (68) in function of the Helmholtz free energy $\hat{\psi}$; precisely, by using $(76)_{5,7}$ in the former and $(76)_{6,8}$ in the latter, we have

$$\left(\operatorname{div}\hat{\mathbf{\Sigma}} - \hat{\mathbf{Z}} \right) \cdot \mathbf{R} = \operatorname{div}\left(\rho\hat{\psi}_{\mathbf{p}} \right) - \rho\hat{\psi}_\beta \quad \text{and} \qquad (82)$$

$$\operatorname{skw}\left[\beta\mathbf{R}\left(\operatorname{div}\hat{\mathbf{\Sigma}} - \hat{\mathbf{Z}} \right)^T \right] = \operatorname{skw}\left\{ \mathbf{R}\left[\operatorname{div}\left(\rho\hat{\psi}_{\mathbf{\Pi}} \right) - \rho\hat{\psi}_{\mathbf{R}} \right]^T \right\},$$

where $\hat{\psi}$ represents a sort of potential for stresses and microstresses.

Now, if we introduce in (82) internal forces of dilatancy δ and of rotation $\mathbf{N}$, the dilating microstress vector $\mathbf{h}_{dil}$ and the third order spinning hyperstress tensor $\mathbf{\Sigma}_{spi}$, defined by

$$\delta := \tfrac{1}{3}\rho\left(\beta\hat{\psi}_\beta + \mathbf{p}\cdot\hat{\psi}_{\mathbf{p}} \right), \quad \mathbf{N} := \tfrac{1}{3}\rho\operatorname{skw}\left(\mathbf{R}\,\hat{\psi}_{\mathbf{R}}^T + \mathbf{\Pi}\otimes\hat{\psi}_{\mathbf{\Pi}} \right), \qquad (83)$$

$$\mathbf{h}_{dil} := \tfrac{1}{3}\rho\beta\hat{\psi}_{\mathbf{p}} \quad \text{and} \quad \mathbf{\Sigma}_{spi}\mathbf{w} := \tfrac{1}{3}\operatorname{skw}\left[\rho\mathbf{R}\left(\hat{\psi}_{\mathbf{\Pi}}\mathbf{w} \right)^T \right], \; \forall\,\text{vector}\,\mathbf{w}, \; (84)$$

as a consequence the balances of dilatational and rotational micromomentum (68) are, respectively:

$$\rho\rho_m \left[\alpha(\rho_m)\left(\ddot{\rho}_m - \rho_m\widetilde{\mathbf{Y}}\cdot\widetilde{\mathbf{Y}} \right) + \tfrac{1}{2}\alpha'(\rho_m)\dot{\rho}_m^2 \right] = \rho\,\phi - \delta + \operatorname{div}\mathbf{h}_{dil}$$

$$\text{and} \quad \rho\left(\rho_m^2\alpha(\rho_m)\widetilde{\mathbf{Y}} \right)^{\textbf{\.{}}} = \rho\,\mathbf{O} - \mathbf{N} + \operatorname{div}\mathbf{\Sigma}_{spi}, \qquad (85)$$

where $\phi := \tfrac{1}{3}\beta\,\mathbf{B}\cdot\mathbf{R}$ and $\mathbf{O} := \tfrac{1}{3}\operatorname{skw}\left(\beta\mathbf{R}\mathbf{B}^T \right)$ are the external dilatational force and the external tensor moment per unit mass, respectively.

These equations for micromomentum together with the balance of mass (18) and the Cauchy's balance of linear momentum (66) ($\widetilde{\mathbf{T}}$ given by (80)) are the pure field equations of motion for thermoelastic dilatant granular materials with rotating grains, the evolution of the temperature being ruled by (81).

Remark 1: We observe now that $\mathbf{h}_{dil}$ and $\boldsymbol{\Sigma}_{spi}$, defined in (84), are particular examples of the stirring and the twisting hyperstress tensor defined in [14] and [8], but, unlike those papers, we think that it is not possible the assignment of prescribed boundary conditions to both of them, the stirrer $\mathbf{h}_{dil}\hat{\mathbf{n}}$ and the twister $\boldsymbol{\Sigma}_{spi}\hat{\mathbf{n}}$ ($\hat{\mathbf{n}}$ is the exterior unit normal to the boundary surface).

In fact they are of different physical nature: while for the twister the boundary distribution of the external couples could be assigned in analogy to the microrigid Cosserat brother's continua [19], on the contrary, for the stirrer, it appears difficult to imagine a direct way to act on the proper grain compressibility through the boundary itself; rather, only the sum $(-\delta + \operatorname{div}\mathbf{h}_{dil})$ has sense, has the right properties of covariance and could express weakly non-local effects (see [7], pages 26–27, [42], page 21, and [5]).

In [11] a wide discussion about the manifold of values of the microstructures with, or without, physically significant connection and the consequent presence, or absence, of the related microstress is presented.

Remark 2: In this context, the mechanical interstitial work flux $\mathbf{u}$, introduced at the end of Sect. 7 in the balance of granular energy (72), is now written as

$$\mathbf{u} = 3\left[\left(\dot{\beta}\beta^{-1}\right)\mathbf{h}_{dil} + \boldsymbol{\Sigma}_{spi}^{T}\widetilde{\mathbf{Y}}\right]; \tag{86}$$

thus terms related to contractions or dilatations of grains and to rotations appear clearly put in evidence.

9 Suspension of Rigid Granules in a Fluid Matrix

In the analysis of flows of a large number of discrete inelastic particles at relatively high concentrations and with interstices filled with a fluid or a gas of negligible mass (as it is the case of cohesionless soil, such as sand with rough surface grains, or of fluidized particulate beds), we must assume that the granules are incompressible; therefore, the proper mass density ρ_m is constant and, for relation (1), the chunk mass density ρ comes down to be proportional to the volume fraction ν of grains ($\rho = \rho_{m*}\nu$) and so the conservation of mass (18) gives

$$\dot{\nu} + \nu\operatorname{tr}\mathbf{L} = 0. \tag{87}$$

Furthermore, for condition $(16)_2$, the coefficient β in the constraint relation (9) disappears ($\beta \equiv 1$) and $\mathbf{G} = \mathbf{R}$, so that

$$\kappa_s = \frac{1}{2}\mu_*^2\dot{\mathbf{R}}\cdot\dot{\mathbf{R}}, \quad \mathbf{W} = \dot{\mathbf{R}}\mathbf{R}^T = \widetilde{\mathbf{Y}}^T, \quad \widetilde{\mathbf{D}} = \mathbf{0} \text{ and } \operatorname{grad}\left(\mathbf{W}^T\right) = \operatorname{grad}\widetilde{\mathbf{Y}}. \tag{88}$$

Remark: When the effects of relative rotations of the chunks and of the grains are also negligible ($\mathbf{R} = \mathbf{I}$), we recover the essence of the theory in [28] and in Sect. 6 of [31]: in particular, in both of them the Coulomb's model for the stress at equilibrium in a granular material with incompressible grains:

$$\mathbf{T}^e = \left(\beta_0 - \beta_1\nu^2 + \beta_2 \operatorname{grad}\nu \cdot \operatorname{grad}\nu + 2\beta_3\,\nu\,\Delta\nu\right)\mathbf{I} - 2\beta_4\operatorname{grad}\nu \otimes \operatorname{grad}\nu$$

with β_i material constants for $i = 0, 1, 2, 3, 4$, is obtained as a peculiar example (see also, (9.1) of [33]). Alternatively, the complementary case in which $\mathbf{R} = \mathbf{I}$, but the grains are elastic, is studied in [30] and again in [31].

We focus here on the simple inelastic case for which relations (88) apply and we develop calculations of Sects. 6–8 with few adjustments.

Firstly, we obtain the following prescriptions for reactions:

$$\left(\mathbf{Z}_r\mathbf{R}^T + \mathbf{\Sigma}_r \odot \operatorname{grad}\mathbf{R}\right) \in \mathrm{Sym}, \quad \mathbf{R}\left(\mathbf{\Sigma}_r\,\mathbf{w}\right)^T \in \mathrm{Sym}, \; \forall\,\text{vector}\,\mathbf{w},$$
$$\mathbf{T}_r = \mathbf{0}, \quad \psi_r = \mathrm{const.}, \quad \eta_r = 0, \quad \mathbf{q}_r = \mathbf{0}, \tag{89}$$

and, correspondingly, for actions

$$\left(\mathbf{Z}_a\mathbf{R}^T + \mathbf{\Sigma}_a \odot \operatorname{grad}\mathbf{R}\right) \in \mathrm{Skw}, \quad \mathbf{R}\left(\mathbf{\Sigma}_a\,\mathbf{w}\right)^T \in \mathrm{Skw}, \; \forall\,\text{vector}\,\mathbf{w},$$
$$\mathbf{T} = \mathrm{sym}\,\mathbf{T}_a - \mathrm{skw}\left(\mathbf{Z}_a\mathbf{R}^T + \mathbf{\Sigma}_a \odot \operatorname{grad}\mathbf{R}\right). \tag{90}$$

Secondly, the reaction-free equation of micromomentum balance for our suspension of rigid granules is now

$$\rho\mu_*^2\dot{\tilde{\mathbf{Y}}} = \mathrm{skw}\left[\mathbf{R}\left(\rho\mathbf{B} - \mathbf{Z}_a + \operatorname{div}\mathbf{\Sigma}_a\right)^T\right], \tag{91}$$

while the equation for the reactions to the constraint is

$$\mathbf{Z}_r - \operatorname{div}\mathbf{\Sigma}_r = \rho\left[\mathrm{sym}\left(\mathbf{B}\,\mathbf{R}^T + \mu_*^2\tilde{\mathbf{Y}}^2\right)\right]\mathbf{R}. \tag{92}$$

We observe that the (91) for the microstructural actions is the same that rules the micromotion for the microrigid Cosserat's continua (see (23.1) of [7] or (63) of [35]).

Thirdly, the set of constitutive variables for a thermoelastic materials with rotating rigid grains is now

$$\mathcal{S}_{rigid} \equiv \{\nu_*, \nu, \operatorname{grad}\nu, \operatorname{grad}^2\nu, \mathbf{R}, \mathbf{\Pi}, \theta_*, \theta, \mathbf{g}\},$$

and so the entropy imbalance (25) and relations (65) and (87) give the following constitutive prescriptions for dependent fields:

$$\rho_{m*}\psi_a = \bar{\psi}(\nu_*, \nu, \mathbf{R}, \mathbf{\Pi}, \theta_*, \theta) =$$
$$= \bar{\psi}\left(\nu_*, \nu, \mathbf{Q}\mathbf{R}, (\mathbf{Q} \oslash \mathbf{\Pi})\,\mathbf{Q}^T, \theta_*, \theta\right), \quad \forall\mathbf{Q} \in \mathrm{Orth}^+,$$

$$\tilde{\mathbf{T}} = -\nu^2\left[\bar{\gamma}(\nu)\ddot{\nu} + \tfrac{1}{2}\bar{\gamma}'(\nu)\dot{\nu}^2 + \bar{\psi}_\nu\right]\mathbf{I} - \nu\mathbf{\Pi}^T \odot \bar{\psi}_{\mathbf{\Pi}}^T, \quad \eta_a = -\bar{\psi}_\theta, \tag{93}$$

$$\mathrm{skw}\left\{\mathbf{R}\left[(\mathbf{\Sigma}_a - \nu\,\bar{\psi}_{\mathbf{\Pi}})\,\mathbf{w}\right]^T\right\} = \mathbf{0}, \quad \forall\,\text{vector}\,\mathbf{w}, \quad \mathbf{q}_a \cdot \mathbf{g} \leq 0,$$

with

$$\bar{\gamma}(\nu) = \bar{\gamma}_* \nu^{-\frac{8}{3}} \quad \text{and} \quad \bar{\gamma}_* = \frac{16}{351} \rho_{m*} \nu_*^{\frac{2}{3}} \varsigma_*^2. \tag{94}$$

In $(93)_3$ the thermodynamic pressure is now $\bar{\pi} := \nu^2 \bar{\psi}_\nu$ and is related to the compressibility of chunks, while the stresses of Reynolds' and of Ericksen's type measure, respectively, the agitation within a chunk of material and the ability of rigid granular continua to support shear stresses in equilibrium, by inducing the generation of microrotation gradients, even when the proper mass density and the volume fraction of the grain distribution is constant.

Finally, the balance of rotational micromomentum (68) is given by

$$\rho_{m*} \nu \mu_*^2 \dot{\tilde{\mathbf{Y}}} = \rho_{m*} \nu \, \bar{\mathbf{O}} - \bar{\mathbf{N}} + \operatorname{div} \bar{\boldsymbol{\Sigma}}_{spi}, \tag{95}$$

where

$$\bar{\mathbf{O}} := \operatorname{skw} \left(\mathbf{R} \mathbf{B}^T \right), \quad \bar{\mathbf{N}} := \nu \operatorname{skw} \left(\mathbf{R} \, \bar{\psi}_{\mathbf{R}}^T + \boldsymbol{\Pi} \otimes \bar{\psi}_{\boldsymbol{\Pi}} \right), \tag{96}$$

$$\bar{\boldsymbol{\Sigma}}_{spi} \mathbf{w} := \operatorname{skw} \left[\nu \mathbf{R} \left(\bar{\psi}_{\boldsymbol{\Pi}} \mathbf{w} \right)^T \right], \quad \forall \, \text{vector} \, \mathbf{w}, \tag{97}$$

are the new external and internal rotational tensor moment per unit mass and the new third order spinning hyperstress tensor, respectively.

The pure field equations of mass, macro- and micromotion and of temperature for granular materials with rigid rotating grains are then (87), (66) with $\widetilde{\mathbf{T}}$ given by $(93)_3$, (95) and (81).

Appendix: Kinetic Energy Coefficients

To compute explicitly the constitutive functions $\gamma(\rho)$ and $\alpha(\rho_m)$ we imagine the chunk consisting, in a mental magnification, of a spherical grain and its immediate spherical neighbours (see [6]), and the envelope of the chunk as a spherical surface of variable radius ς containing all these spherical compressible inclusions of variable radius φ with interstices filled with a fluid or a gas of negligible mass; the envelopes and the grains have the same radius ς_* and φ_*, respectively, in a reference placement $\mathcal{B}_*$ of the material.

Moreover, we assume that the chunks and the grains expand and/or contract homogeneously with independent radial motions; therefore, if we indicate with $\tilde{\varsigma}$ the distance from the centre of mass of the chunk to the centre of mass of a grain in the chunk itself, and with $\tilde{\varphi}$ the distance from the centre of mass of a grain to the element of volume dv_m, they are related to ς and φ by

$$\tilde{\varsigma} = \frac{\tilde{\varsigma}_*}{\varsigma_*} \varsigma \quad \text{and} \quad \tilde{\varphi} = \frac{\tilde{\varphi}_*}{\varphi_*} \varphi, \tag{98}$$

respectively.

Furthermore, the average density of kinetic energy $(\kappa_f^{ch} + \kappa_d^{ch})$ per unit volume associated to each chunk, as effect of the homogeneous expansions or contractions of a typical chunk itself and of the inclusions (in addition to the classical kinetic energy of translation κ_t^{ch}), will be written

$$\kappa_f^{ch} + \kappa_d^{ch} = \frac{1}{2} \frac{1}{\frac{4}{3}\pi\varsigma^3} \sum^n \left(m\dot{\varsigma}^2 + \int_{\mathcal{V}_m} \rho_m \dot{\varphi}^2 \, d\mathcal{V}_m \right), \qquad (99)$$

where $\sum$ denotes summation over all of the grains of the chunk, n is the number of the grains in a chunk, $\mathcal{V}_m$ and m are the volume and the mass of a typical grain, respectively, i.e.,

$$\mathcal{V}_m = \frac{4}{3}\pi\varphi^3 \quad \text{and} \quad m = \frac{4}{3}\rho_m\pi\varphi^3 = \frac{4}{3}\rho_{m*}\pi\varphi_*^3 \qquad (100)$$

(see also, (2.3) of [4]); hence, from relations $(98)_2$ and $(100)_3$, the time rate of change of $\tilde{\varphi}$ can be expressed in terms of the rate of change of ρ_m:

$$\dot{\tilde{\varphi}} = -\frac{1}{3}\tilde{\varphi}_* \, \rho_{m*}^{\frac{1}{3}} \, \rho_m^{-\frac{4}{3}} \, \dot{\rho}_m. \qquad (101)$$

We observe that the quasi-particles are assumed to fill the space of the granular material, without voids between them, and so the volume fraction ν of the chunk is

$$\nu = \frac{1}{\frac{4}{3}\pi\varsigma^3} \sum^n \frac{4}{3}\pi\varphi^3 = \sum^n \left(\frac{\varphi}{\varsigma}\right)^3, \qquad (102)$$

while (being ς_* and φ_* constants in the same chunk)

$$\nu_* = \sum^{n_*} \left(\frac{\varphi_*}{\varsigma_*}\right)^3 = n_* \left(\frac{\varphi_*}{\varsigma_*}\right)^3. \qquad (103)$$

Moreover, the granules are supposed homogeneous, strictly packed and such that they do not diffuse throughout the envelope of the chunk; therefore, $\rho_{m*} = \text{const.}$, the immediate neighbours of a grain are twelve, with $n = n_* = 13$, and, finally, the volume and the mass density of a macroelement are, respectively,

$$\mathcal{V} = \frac{4}{3}\pi\varsigma^3 \quad \text{and} \quad \rho = \frac{\rho_* \mathcal{V}_*}{\mathcal{V}} = \rho_* \left(\frac{\varsigma_*}{\varsigma}\right)^3. \qquad (104)$$

Thus, it follows from $(98)_1$ and $(104)_3$ that the time rate of change of $\tilde{\varsigma}$ can be expressed in terms of the rate of change of ρ

$$\dot{\tilde{\varsigma}} = -\frac{1}{3}\tilde{\varsigma}_* \, \rho_*^{\frac{1}{3}} \, \rho^{-\frac{4}{3}} \, \dot{\rho} \qquad (105)$$

and, from (100)$_3$, (104)$_3$ and (105), that the 'fluctuation' kinetic energy κ_f^{ch} of the chunk is

$$\kappa_f^{ch} = \frac{1}{2}\frac{1}{\frac{4}{3}\pi\varsigma^3}\sum^n m\dot{\tilde{\varsigma}}^2 = \frac{\rho_*^{\frac{2}{3}}\dot{\rho}^2}{18\rho^{\frac{8}{3}}\varsigma^3}\sum^{n_*}\rho_{m*}\varphi_*^3\tilde{\varsigma}_*^2 = \frac{\rho_{m*}\varphi_*^3\dot{\rho}^2}{18\varsigma_*^3\rho_*^{\frac{1}{3}}\rho^{\frac{5}{3}}}\sum^{n_*}\tilde{\varsigma}_*^2, \quad (106)$$

where we used the fact that the single grains of a chunk are homogeneous and of the same radius φ_* in the reference placement $\mathcal{B}_*$ of the material.

Nevertheless, we supposed the granules strictly packed in $\mathcal{B}_*$, therefore, the centre of mass of the chunk coincides with the centre of the main grain (hence the related $\tilde{\varsigma}_*$ vanishes), while the centre of mass of its twelve immediate spherical neighbours are distant two time the constant radius φ_* of a grain from the centre of the main grain (hence, $\tilde{\varsigma}_* = 2\varphi_*$); moreover, the radius of the chunk envelope ς_* is three time the radius φ_*, i.e., $\varsigma_* = 3\varphi_*$ and $\tilde{\varsigma}_* = \frac{2}{3}\varsigma_*$.

Thus, by using also relations (1) and (103), we have

$$\rho_{m*}\left(\frac{\varphi_*}{\varsigma_*}\right)^3 = \frac{\rho_*}{n_*} \quad \text{and} \quad \sum\tilde{\varsigma}_*^2 = \frac{4}{9}(n_* - 1)\varsigma_*^2; \quad (107)$$

at the end, by inserting (107) and $n_* = 13$ in (106)$_3$, we obtain the density of kinetic energy κ_f^{ch} per unit volume associated to each chunk

$$\kappa_f^{ch} = \frac{1}{2}\rho\left(\frac{16}{351}\rho_*^{\frac{2}{3}}\varsigma_*^2\right)\rho^{-\frac{8}{3}}\dot{\rho}^2. \quad (108)$$

Now, if we consider the 'dilatational' kinetic energy κ_d^{ch} of the chunk and use relations (100), (101), (104)$_3$ and, after, (1) and (103)$_1$, we obtain

$$\kappa_d^{ch} = \frac{1}{2}\frac{1}{\frac{4}{3}\pi\varsigma^3}\sum^n\int_{\mathcal{V}_m}\rho_m\dot{\tilde{\varphi}}^2\,d\mathcal{V}_m = \frac{\rho\rho_{m*}^{\frac{5}{3}}\dot{\rho}_m^2}{24\pi\rho_*\varsigma_*^3\rho_m^{\frac{8}{3}}}\sum^{n_*}\int_0^{\varphi_*}4\pi\tilde{\varphi}_*^4\,d\tilde{\varphi}_* =$$

$$= \frac{\rho\rho_{m*}^{\frac{2}{3}}\varphi_*^2\dot{\rho}_m^2}{30\rho_*\varsigma_*^3\rho_m^{\frac{8}{3}}}\left[\frac{\rho_{m*}}{\rho_*}\sum^{n_*}\left(\frac{\varphi_*}{\varsigma_*}\right)^3\right] = \frac{1}{2}\rho\left(\frac{1}{15}\rho_{m*}^{\frac{2}{3}}\varphi_*^2\right)\rho_m^{-\frac{8}{3}}\dot{\rho}_m^2. \quad (109)$$

The constitutive expressions (6) and (7) for the coefficients $\gamma(\rho)$ and $\alpha(\rho_m)$, which appear in formula (5) are then easily recognized in (108) and (109)$_4$.

Acknowledgement

This research was supported by the Italian "Gruppo Nazionale di Fisica Matematica" of the "Istituto Nazionale di Alta Matematica" and by the Italian M.U.R.S.T. through the project "PRIN2005 – Modelli Matematici per la Scienza dei Materiali". The author is also grateful to P.M. Mariano for some helpful discussion.

References

1. Ahmadi, G.: A Generalized Continuum Theory for Granular Materials. Int. J. Non–Linear Mech., **17**, 21–33 (1982)
2. Arthur, J.R.F., Menzies, B.K.: Inherent Anisotropy in Sand. Geotechnique, **22**, 115–129 (1972)
3. Bampi, F., Morro, A.: The Connection between Variational Principles in Eulerian and Lagrangian Descriptions. J. Math. Phys., **25**, 2418–1421 (1984)
4. Bedford, A., Drumheller, D.S.: On Volume Fraction Theories for Discretized Materials. Acta Mechanica, **48**, 173–184 (1983)
5. Bridgman, P.W.: Reflections on Thermodynamics. Proc. Am. Acad. Arts Sci., **82**, 301–309 (1953)
6. Buyevich, Y.A., Shchelchkova, I.N.: Flow of Dense Suspensions. Prog. Aerospace Sci., **18**, 121–150 (1978)
7. Capriz, G.: Continua with Microstructure. Springer Tracts in Natural Philosophy. Springer-Verlag, Berlin Heidelberg New York, **35**, (1989)
8. Capriz, G.: Pseudofluids. In: Capriz, G., Mariano, P.M. (eds) Material Substructures in Complex Bodies: from Atomic Level to Continuum. Elsevier Science B.V., Amsterdam, 238–261 (2006)
9. Capriz, G.: Elementary Preamble to a Theory of Granular Gases. Rend. Semin. Mat. Padova, **110**, 179–198 (2003)
10. Capriz, G., Giovine, P.: On Microstructural Inertia. Math. Mod. Meth. Appl. Sc., **7**, 211–216 (1997)
11. Capriz, G., Giovine, P.: Remedy to Omissions in a Tract on Continua with Microstructure. Atti XIII Congresso AIMETA '97, Siena. Meccanica Generale, **I**, 1–6 (1997)
12. Capriz, G., Giovine, P.: Weakly Nonlocal Effects in Mechanics. In: Contributions to Continuum Theories, Krzysztof Wilmanski's Anniversary Volume, Weierstrass Institute for Applied Analysis and Stochastics, Berlin, Report n.18 - ISSN 0946-8838, 37–44 (2000)
13. Capriz, G., Mullenger, G.: Extended Continuum Mechanics for the Study of Granular Flows. Rend. Acc. Lincei, Matematica, **6**, 275–284 (1995)
14. Capriz, G., Mullenger, G.: Dynamics of Granular Fluids. Rend. Sem. Mat. Univ. Padova, **111**, 247–264 (2004)
15. Capriz, G., Podio–Guidugli, P.: Formal Structure and Classification of Theories of Oriented Materials. Annali Mat. Pura Appl. (IV), **CXV**, 17–39 (1977)
16. Capriz, G., Podio–Guidugli, P.: Materials with Spherical Structure. Arch. Rational Mech. Anal., **75**, 269–279 (1981)
17. Capriz, G., Podio–Guidugli P., Williams, W.: Balance Equations for Materials with Affine Structure. Meccanica, **17**, 80–84 (1982)
18. Capriz, G., Podio–Guidugli, P.: Internal Constraint. In: Truesdell, C.: Rational Thermodynamics (second edition), Springer-Verlag, New York, Appendix **3A**, 159–170 (1984)
19. Cosserat, E. and F.: Théorie des Corps Déformables. Hermann (1909)
20. Cowin, S.C.: A Theory for the Flow of Granular Materials. Powder Technology, **9**, 61–69 (1974)
21. Cowin, S.C.: Microstructural Continuum Models for Granular Materials. In: Cowin, S.C., Satake, M. (eds) Proc. USA–Japan Seminar on Continuum–Mechanical and Statistical Approaches in the Mechanics of Granular Materials, Tokyo. Gakujutsu Bunken Fukyu–Kai (1978)

22. Cowin, S.C., Nunziato, J.W.: Waves of Dilatancy in a Granular Material with Incompressible Granules. Int. J. Engng. Sci., **19**, 993–1008 (1981)
23. Dell'Isola, F., Seppecher, P.: Edge Contact Forces and Quasi–Balanced Power. Meccanica, **32**, 33–52 (1997)
24. Dunn, J.E., Serrin, J.: On the Thermomechanics of Interstitial Working. Arch. Rat. Mech. Anal., **88**, 95–133 (1985)
25. Dunn, J.E.: Interstitial Working and a Nonclassical Continuum Thermodynamics. In: Serrin, J. (ed) New Perspectives in Thermodynamics. Springer–Verlag, Berlin, 187–222 (1986)
26. Eringen, A.C.: Mechanics of Micromorphic Continua. In: Kröner, E. (ed) Proc. IUTAM Symposium on Mechanics of Generalized Continua, Freudenstadt and Stuttgart (1967) Springer, Berlin Heidelberg New York, 18–35 (1968)
27. Giovine, P.: Termodinamica dei Continui Granulari Dilatanti. Atti XII Congresso AIMETA '95, Napoli. Meccanica dei Fluidi, **IV**, 287–292 (1995)
28. Giovine, P.: Nonclassical Thermomechanics of Granular Materials. Mathematical Physics, Analysis and Geometry, **2**, 179–196 (1999)
29. Giovine, P., Mullenger, G., Oliveri, F.: Remarks on Equations for Fast Granular Flow. In: Inan, E., Markov, K. (eds) Continuum Models and Discrete Systems. World Scientific Publishing Co., Singapore, **IX**, 787–794 (1998)
30. Giovine, P., Oliveri, F.: Dynamics and Wave Propagation in Dilatant Granular Materials. Meccanica, **30**, 341–357 (1995)
31. Giovine, P., Speciale, M.P.: On Interstitial Working in Granular Continuous Media. In: Ciancio, V., Donato, A., Oliveri, F., Rionero, S. (eds) Proceed. 10th Int. Conf. on Waves and Stability in Continuous Media (WASCOM'99), Vulcano (Me). World Scientific, Singapore, 196–208 (2001)
32. Godano, C., Oliveri, F.: Nonlinear seismic waves: a model for site effects. Int. J. Non-linear Mech., **34**, 457–468 (1999)
33. Goodman, M.A., Cowin, S.C.: A Continuum Theory for Granular Materials. Arch. Rational Mech. An., **44**, 249–266 (1972)
34. Goodman, M.A., Cowin, S.C.: A Variational Principle for Granular Materials. ZAMM, **56**, 281–286 (1976)
35. Grioli, G.: Microstructures as a Refinement of Cauchy Theory. Problems of Physical Concreteness. Continuum Mech. Thermodyn., **15**, 441–450 (2003)
36. Gurtin, M.E., Podio–Guidugli, P.: The Thermodynamics of Constrained Materials. Arch. Rational Mech. Anal., **51**, 192–208 (1973)
37. Hutter, K., Rajagopal, K.R.: On Flows of Granular Materials. Continuum Mech. Thermodyn., **6**, 81–139 (1994)
38. Jenkins, J.T.: Static Equilibrium of Granular Materials. J. Appd. Mech., **42**, 603–606 (1975)
39. Johnson, P.C., Jackson, R.: Frictional–Collisional Constitutive Relations for Granular Materials, with Applications to Plane Shearing. J. Fluid Mech., **176**, 67–93 (1987)
40. Kanatani, K.I.: A Micropolar Continuum Theory for the Flow of Granular Materials. Int. J. Engng. Sci., **17**, 419–432 (1979)
41. Ko, H.Y., Scott, R.F.: J. Soil Mech. and Found. Div., Proc. ASCE **93**, 137–156 (1967)
42. Kunin, I.A.: Elastic Media with Microstructure II. Springer Series in Solid–State Sciences, **44**, Springer–Verlag, Berlin (1983)
43. Mariano, P.M.: Multifield Theories in Mechanics of Solids. Advances in Applied Mechanics, **38**, 1–93 (2002)

44. Mitarai, N., Hayakawa, H., Nakanishi, H.: Collisional Granular Flow as a Micropolar Fluid. Phys. Rev. Lett., **88**, 174301 (2002)

45. Nunziato, J.W., Walsh, E.K.: One-Dimensional Shock Waves in Uniformly Distributed Granular Materials. Int. J. Solids Structures, **14**, 681–689 (1978)

46. Passman, S.L.: Mixtures of Granular Materials. Int. J. Engng. Sci., **15**, 117–129 (1977)

47. Passman, S.L., Nunziato, J.W., Bailey, P.B., Thomas jr., J.P.: Shearing Flows of Granular Materials. J. Engng. Mech. Div. ASCE, **4**, 773–783 (1980)

48. Passman, S.L., Nunziato, J.W., Walsh, E.K.: A Theory of Multiphase Mixtures. In: Truesdell, C., Rational Thermodynamics. Springer-Verlag, New York, 286–325 (1984)

49. Reynolds, O.: On the Dilatancy of Media Composed of Rigid Particles in Contact. Phil. Magazine, **20**, 469–481 (1885)

50. Savage, S.B.: Gravity Flow of Cohesionless Granular Materials in Chutes and Channels. J. Fluid Mech., **92**, part 1, 53–96 (1979)

51. Truesdell, C., Toupin, R.A.: The Classical Field Theories. Handbuch der Physik, **III/1**, Flügge, S. (ed), Springer–Verlag, Berlin (1960)

Slow Motion in Granular Matter

Paolo Maria Mariano

Summary. Agglomerations of granules are described as continuous complex bodies in which the generic material element is an open system made of a family of granules. Inertia is neglected while migration of granules is allowed: an evolution equation for the local numerosity of granules is derived in the present setting. In a reduced framework in which the generic material element coincides with a single granule, the balance of interactions governing the motion of a single granule during segregation is also discussed.

1 Introduction

Granular matter is said to be in *slow motion* when inertia effects and effects due to collisions between granules are negligible and only 'slow' migration (or better segregation) of granules, clustering with respect to sizes, occurs. As a consequence, besides equilibrium conditions, only the evolution equation of the density of granules has to be determined, especially when more than one family of granules is present. Such a type of equation is derived here in a continuum representation of granular matter in which I consider each material element as made of a family of granules each of them being free to migrate toward the neighboring material elements (each one being the smallest patch of matter characterizing the body). The path followed is strictly the one developed in [18] for general complex bodies allowing migration of substructures between material elements, elements that are then considered as open systems. The basic reason is that a granular material can in principle be considered as a complex body, a body in which the material texture (here the manner in which granules are arranged in clusters) influences prominently the gross behavior in a way in which the interactions due to the changes of the texture itself cannot be neglected and must be represented directly.

The physical manifestations of granular agglomerations in fast and slow motions are manifold (see remarks in [1], [14], [15], [16]) and the appearance of entangled effects suggest different theoretical approaches that may describe some aspects and neglect others (as it is natural). Fast motion of granular matter suggests microscopic approaches based on (inelastic) Boltzmann equation (see [2], [3], [8], [11]). The description of equilibrium states is often based on points of view arising from the mechanics of complex bodies (see, e.g., [12], [21]) or from the direct statistical analyses of contact forces amid granules [9]. In the middle there is an approach proposed in [6], [7], an approach in which hydrodynamic equations are obtained from first principles by analyzing directly the motion of single granules within the assembly contained by the generic material element.

The approach proposed here is reminiscent of remarks in [6], [12] and [9]; however, differences are evident, above all in the treatment of the evolution of the numerosity.

The case in which each material element is made of a single granule undergoing own rotations is treated last: the force driving a single granule with respect to the neighboring fellows is then deduced.

2 Representation of the Granularity

A body may be considered as an abstract set $\mathfrak{B}$, each $\mathfrak{e} \in \mathfrak{B}$ being the smallest piece of matter characterizing the material composing it, i.e. a 'representative volume element', called also *material element*, in the common parlance. The essential starting point is then the representation of $\mathfrak{B}$, precisely the geometrical structure one attributes to $\mathfrak{B}$ by mapping it in some other set. In standard continuum mechanics, each $\mathfrak{e}$ is considered as a windowless box described only by the place in space occupied by its centre of mass [23], [22]. However, the effects of the changes in the structure internal to $\mathfrak{e}$, the *material substructure*, cannot be often neglected because interactions conjugated with these changes determine prominent effects. In this case bodies are called complex. In the mechanics of complex bodies (see [5], [17], [20] for general issues) the basic view is to assign to each material element a morphological descriptor of the inner substructure, a descriptor that is selected in general as an element of a finite-dimensional differentiable manifold, which is then the *manifold of substructural shapes*. In this way the description of the body becomes multifield since one manages the placement field of $\mathfrak{B}$ and the field of morphological descriptors of its substructure.

I adopt this point of view here with the aim of constructing a model of granular matter in the continuum limit. For such a purpose, the use of two isomorphic copies of $\mathbb{R}^3$, indicated respectively by $\mathbb{R}^3$ and $\hat{\mathbb{R}}^3$, $\iota : \hat{\mathbb{R}}^3 \to \mathbb{R}^3$ the isomorphism, is needed. In $\mathbb{R}^3$ the generic element $\mathfrak{e}$ of $\mathfrak{B}$ is represented by a point while $\hat{\mathbb{R}}^3$ is used to describe the granularity *within* $\mathfrak{e}$. In this sense, $\mathbb{R}^3$ and $\hat{\mathbb{R}}^3$ collect the events occurring at two scales. In $\hat{\mathbb{R}}^3$ a *prototype material element* $\mathfrak{e}$ is expanded. One may consider $\mathfrak{e}$ (i) as composed only by a single granule or (ii) as made of a family of $\hat{n}$ cohesionless granules each one placed at $\hat{x}_r$, $r = 1, ..., \hat{n}$. All granules have the same mass $\hat{m}$. In both cases (*i*) and (*ii*) no attention is paid to the shape of granules. Contact between adjacent granules is with perfect friction.

The *total mass* m of $\mathfrak{e}$ is then

$$m = \hat{n}\hat{m}.$$

The position $\hat{x}$ of the *centre of mass of* $\mathfrak{e}$ is given by

$$\hat{x} := \frac{1}{\hat{n}} \sum_{r=1}^{n} \hat{x}_r. \tag{1}$$

After introducing the *relative position vector*

$$\hat{z}_r := \hat{x}_r - \hat{x}, \tag{2}$$

with $\sum_{r=1}^{n} \hat{z}_r = 0$, a natural descriptor of the shape of the family of granules within $\mathfrak{e}$ is the second rank symmetric tensor

$$Y := \frac{1}{\hat{n}} \sum_{r=1}^{n} \hat{z}_r \otimes \hat{z}_r.$$

Its trace indicates an average of the squared distance of the granules from $\hat{x}$ (see [10]). $Y \cdot (\hat{r} \otimes \hat{r}) \geq 0$ for any vector $\hat{r} \in \hat{\mathbb{R}}^3$; also Y is symmetric by definition. Essentially it is the moment of inertia of the family of granules.

Functions $[0, \bar{t}] \ni t \mapsto \hat{x}_r(t)$, $r = 1, ..., \hat{n}$, describe in $\hat{\mathbb{R}}^3$ the motion of each granule. As a consequence, the *linear momentum* $\hat{p}$ of the system of granules within $\mathfrak{e}$ is given by

$$\hat{p} := \sum_{r=1}^{n} \hat{m} \frac{d\hat{x}_r}{dt} = m \frac{d\hat{x}}{dt},$$

when the numerosity of granules within $\mathfrak{e}$ remains constant in time. The *tensor moment of momentum* $\hat{\mathfrak{m}}$ of the same system is defined by

$$\hat{\mathfrak{m}} := \sum_{r=1}^{n} \hat{m} \left(\hat{x}_r \otimes \frac{d\hat{x}_r}{dt} - \frac{d\hat{x}_r}{dt} \otimes \hat{x}_r \right). \tag{3}$$

By using (1) and (2), elementary algebra allows one to rewrite (3) in the form

$$\hat{\mathfrak{m}} = 2 \left(\sum_{r=1}^{n} \hat{m}\hat{z}_r \otimes \frac{d\hat{z}_r}{dt} + skw\,(\hat{x} \otimes \hat{p}) \right) - m\dot{Y}, \tag{4}$$

where skw extract the antisymmetric part of its argument. Notice that the rate of Y describes only the possible shuffling of the material element, not the rotational effects of the granules inside it.

A crucial step below is the identification (through the isomorphism) of the place $\hat{x}$ in $\hat{\mathbb{R}}^3$ of the centre of mass of the system of granules with the place $x := \iota\left(\hat{x}\right)$ assigned in $\mathbb{R}^3$ to the whole material element $\mathfrak{e}$. Then one may identify the linear momentum $\hat{p}$ of the family of granules within $\mathfrak{e}$ with the linear momentum $p := m\frac{dy(x,t)}{dt}$ of the mass point $y(x,t)$ in $\mathbb{R}^3$ corresponding to x when $t = 0$ and to $\iota\left(\hat{x}\left(t\right)\right)$ at any t. Consequently, the relation (4) suggests that the intrinsic tensor moment of momentum of each material element is given by the sum of two contributions: .

(a) the moment of momentum of the entire system, namely $2skw\left(\hat{x} \otimes \hat{p}\right)$, a quantity that can be identified with the tensor moment of momentum of the mass point at y in $\mathbb{R}^3$, which is $2skw\left(y \otimes p\right)$, and

(b) the moment of momentum of the *agitation* within $\mathfrak{e}$:

$$2\sum_{r=1}^{n} \hat{m}\hat{z}_r \otimes \frac{d\hat{z}_r}{dt} - m\dot{Y} = \sum_{r=1}^{n} \hat{m}\left(\hat{z}_r \otimes \frac{d\hat{z}_r}{dt} - \frac{d\hat{z}_r}{dt} \otimes \hat{z}_r\right).$$

The collection of all centres of mass of all elements of $\mathfrak{B}$, identified with points in $\mathbb{R}^3$, is a *place* (or better a *gross place*) of the body. It is defined by the image of a one-to-one map

$$k_p : \mathfrak{B} \to \mathbb{R}^3.$$

The set $\mathcal{B} := k_p\left(\mathfrak{B}\right)$ is assumed to be a bounded domain of $\mathbb{R}^3$ with boundary $\partial\mathcal{B}$ of finite two-dimensional measure, a boundary where the outward unit normal n is defined to within a finite number of corners and edges. It is convenient to select a reference place $\mathcal{B}_0 = k_{p,0}\left(\mathfrak{B}\right)$, the generic point of which is labeled by x. All other places, the generic one being indicated by $\mathcal{B}$, are achieved from $\mathcal{B}_0$ by means of a *standard transplacement* y which maps $\mathcal{B}_0$ onto $\mathcal{B}$, namely

$$\mathcal{B}_0 \ni x \mapsto y\left(x\right) := \left(k_p \circ k_{p,0}^{-1}\right)\left(x\right) \in \mathcal{B}.$$

As usual, the transplacement map y is assumed to be (i) *one-to-one*, (ii) *piecewise continuously differentiable* and (iii) *orientation preserving*. The latter requirement implies that at each x the spatial derivative of y, a derivative indicated by F, and called as usual *gradient of deformation*, has positive determinant at each point. F is the value of the map

$$\mathcal{B}_0 \ni x \mapsto F := F\left(x\right) \in Hom\left(T_x\mathcal{B}_0, T_{y(x)}\mathcal{B}\right) \simeq \mathbb{R}^3 \otimes \mathbb{R}^3 = M_{3\times3}$$

such that $\det F > 0$. In deforming the body from $\mathcal{B}_0$ to $\mathcal{B}$, the material elements crowd and\or shear with each other. A measure of these mechanisms, that is a measure of how lengths and angles in $\mathcal{B}_0$ change as a consequence of the deformation, is easily available when one assign metrics in $\mathcal{B}_0$ and $\mathcal{B}$, say γ and g respectively, and compare them in a common paragon setting. For example, if one decides that the appropriate paragon setting is $\mathcal{B}_0$, by pulling back g in $\mathcal{B}_0$ by means of y, and writing $y^{\#}g$ for the pull-back,

$$y^{\#}g := F^*F \in Hom\,(T_x\mathcal{B}_0, T_x^*\mathcal{B}_0)\,,$$

half of the difference $y^{\#}g - \gamma$ defines the non-linear deformation tensor E that is the desired pointwise measure of gross deformation. In the common notation of textbooks in continuum mechanics, the linear operator F^*F, the right Cauchy–Green tensor, is indicated by C, in components $C_{AB} = F_A^{*i}g_{ij}F_B^j$, so that E is defined by $E := \frac{1}{2}\,(C - \gamma)$.

When one assigns a place to a material element $\mathfrak{e}$, the distances inside $\mathfrak{e}$, say the dimensions of the granules and/or the distances between the centres of mass of neighboring granules, become *internal* lengths because $\mathfrak{e}$ is collapsed in a point. This is the point of view of standard continuum mechanics of Cauchy's bodies. No information about the granularity inside $\mathfrak{e}$ is furnished by the placement map k_p, so that the representation of $\mathfrak{B}$ is not sufficient. At least coarse grained information about the granularity have to be accounted for at 'kinematical' level. They can be introduced by means of a *morphological descriptor* which summarizes the main geometrical information about the family of granules within the generic $\mathfrak{e}$. In this way information in $\hat{\mathbb{R}}^3$ (where the material element is expanded) about the discrete system of granules can be translated in a continuous field theory in $\mathbb{R}^3$. This is the basic reason for resorting to two different isomorphic copies of $\mathbb{R}^3$.

In all relations written above in $\hat{\mathbb{R}}^3$, it is implicitly presumed that one knows the exact place $\hat{x}_r$ of each granule within each material element $\mathfrak{e} \in \mathfrak{B}$ when it is expanded in $\hat{\mathbb{R}}^3$. Such an assumption is of course highly optimistic because at most one knows a distribution $\theta\,(\hat{x}', \hat{v})$ of places and velocities (see [7]). However, since only slow motion is under scrutiny here, it is sufficient to make use of a reduced distribution $\hat{\theta}\,(\hat{x}')$ of sole places, a distribution coinciding with the average of θ over the space of velocities.

Quantities in $\hat{\mathbb{R}}^3$ pertinent to $\mathfrak{e}$ may be then re-defined in terms of $\hat{\theta}$ so that the centre of mass $\hat{x}$ is given by

$$\hat{x} := \frac{1}{n}\int_{\hat{\mathbb{R}}^3} \hat{\theta}\,(\hat{x}')\,\hat{x}'\,d\hat{x}',$$

and, after choosing the local frame in $\hat{\mathbb{R}}^3$ in a way in which $\hat{x} = 0$, the shape tensor $\hat{Y}$ (really the moment of inertia) can be defined by

$$\hat{Y} := \int_{\hat{\mathbb{R}}^3} \hat{\theta}\,(\hat{x}')\,\hat{x}' \otimes \hat{x}'\,d\hat{x}',$$

where the normalization condition $\frac{1}{n}\int_{\hat{\mathbb{R}}^3} \hat{\theta}\,(\hat{x}')\,d\hat{x}' = 1$ is assumed to hold. $\hat{Y}$ is the counterpart of Y, the latter tensor defined by considering $\mathfrak{e}$ as a deterministic (discrete) system. A *characteristic internal length* ℓ pertinent to the material element can be then defined by

$$\ell := \left(\int_{\hat{\mathbb{R}}^3} \hat{\theta}\,(\hat{x}')\,|\hat{x}'|^2\,d\hat{x}'\right)^{\frac{1}{2}}.$$

Take note that when the number of granules changes within the material element, one is forced to consider a family $\hat{\theta}_t$ of distributions parametrized by time. If one assumes that the material element occupies in $\hat{\mathbb{R}}^3$ a compact domain with finite volume, a domain indicated also by $\mathfrak{e}$, then $\hat{\theta}$ has compact support and the integrals above over the entire $\hat{\mathbb{R}}^3$ reduce simply to integrals over $\mathfrak{e}$.

All the elements above concur in suggesting to select a second-rank tensor ν as appropriate *morphological descriptor* of the granularity within the generic material element, namely

$$\nu \in Hom\left(\mathbb{R}^3, \mathbb{R}^3\right).$$

The space $Hom\left(\mathbb{R}^3, \mathbb{R}^3\right)$ plays here the role of manifold of substructural shapes, the collection of possible inner shapes of the material element, a manifold indicated below by $\mathcal{M}$.

In addition to the placement map k_p, another map

$$k_m : \mathfrak{B} \to \mathcal{M}$$

assigns then to each $\mathfrak{e} \in \mathfrak{B}$ the morphological descriptor of the granular structure inside it. One then gets

$$\mathcal{B}_0 \ni x \mapsto \nu = \nu\left(x\right) := \left(k_m \circ k_{p,0}^{-1}\right)\left(x\right) \in \mathcal{M}, \qquad (5)$$

a map assumed here differentiable. Its spatial derivative is indicated by N so that, at each x,

$$N := D\nu\left(x\right) \in Hom\left(T_x\mathcal{B}_0, T_{\nu(x)}\mathcal{M}\right).$$

By exploiting the reference place $\mathcal{B}_0$ for the Lagrangian (referential) description of fields, along a given interval of time $[0, \bar{t}]$, motions are then time parametrized mappings of places and morphological descriptors so that one has time–space fields

$$\mathcal{B}_0 \times [0, \bar{t}] \ni (x, t) \mapsto y = y\left(x, t\right) \in \mathcal{B},$$

$$\mathcal{B}_0 \times [0, \bar{t}] \ni (x, t) \mapsto \nu = \nu\left(x, t\right) \in \mathcal{M}.$$

In this coarse grained representation, direct information about the numerosity of granules is now given by a map

$$\phi : \mathcal{M} \to \mathbb{R}^+,$$

a distribution over $\mathcal{M}$ assigning to the generic material element, with inner granular arrangement ν, the number of granules involved in that arrangement (strictly, in the actual coarse grained representation, ϕ takes the role of $\hat{n}$ used in $\hat{\mathbb{R}}^3$). The explicit assignment of ϕ at the initial instant is a *constitutive prescription.* The numerosity can be considered as a field over $\mathcal{B}_0$ *along* ν when one introduces the map

$$\alpha := \phi \circ \nu : \mathcal{B}_0 \to \mathbb{R}^+,$$

that will be of later use. α is assumed to be of class C^1.

At a given place, time variations of ν may be determined by (i) changes in numerosity of granules, (ii) variations of the characteristic internal length and (iii) rearrangement of granules. Each material element is then an *open system* that may exchange both energy (by contact interactions) and mass (through migration of granules) with the neighboring fellows. An interpretation of the physical meaning to be attributed to ν accrues by considering once more $\mathbf{e}$ expanded in $\hat{\mathbb{R}}^3$ and assuming that each granule be characterized by an affine motion in $\hat{\mathbb{R}}^3$ that is, to within an additive constant vector,

$$\frac{d\hat{x}_r}{dt} = \hat{A}\hat{x}_r,$$

with $\hat{A}$ a 3×3 matrix independent of r. The independence of r is tantamount to assume that the system of granules inside $\mathbf{e}$ suffers a *global* affine motion. By polar decomposition, one would then get uniquely $\hat{A} = \hat{R}\hat{U}$, with $\hat{R}$ and $\hat{U}$ orthogonal and symmetric tensors respectively. ν has then the meaning of $\hat{A}$. It accounts for both local rotations of granules (the ones described by $\hat{R}$ in $\hat{\mathbb{R}}^3$) and the shuffling of granules themselves (such an effect is accounted for by $\hat{U}$). If $\hat{A}$, then ν, coincides with $\hat{U}$, the effect of the local rotations is neglected. In particular, when ν belongs to the subspace of $\mathcal{M}$

$$\left\{ \nu \in Hom\left(\mathbb{R}^3, \mathbb{R}^3\right) \mid \nu = \nu^*, \ \nu \cdot (r \otimes r) > 0, \ \forall r \in \hat{\mathbb{R}}^3, \ \hat{r} \neq 0 \right\}$$

then it can be identified with $\hat{Y}$ that one may normalize further by dividing it by ℓ^2. The requirement

$$\nu \cdot (r \otimes r) > 0, \ \forall r \in \hat{\mathbb{R}}^3, \ \hat{r} \neq 0$$

implies that it is excluded that the centres of mass of all granules within $\mathbf{e}$ lie on a plane or along a line. When $\hat{A}$ coincides only with $\hat{R}$, only the effect of local rotations inside $\mathbf{e}$ are accounted for. The same interpretation holds when ν coincides with $\hat{R}$.

The model presented here is then the one of a body with affine structure, differences with respect to standard continuum schemes of affine bodies (see, e.g., [5]) relying above all in the assumption that the material element is an open system, so that (contrary to standard issues) the elements of the internal structure (here the granules) may migrate from one material element to another. The other essential difference is that I consider a distribution of the numerosity of granules by adding so information on the material substructure (the general structure in [18] is adopted this way). Due to the link between

α and ν, I consider below only substructural interactions associated with the rate of change of ν, in addition to the standard ones power conjugated with the rate of change of place. It is then assumed that the substructural interactions selected include also the ones associated with the rate of change of numerosity. Consequently, it is only necessary that α satisfies a continuity equation that is intrinsically the balance of mass in the present setting.

3 Balance of Interactions: $\mathbb{R}^3 \ltimes SO(3)$ Invariance

At each pair (x, t), time rates are indicated by

$$\dot{y} := \frac{dy(x,t)}{dt}, \quad \dot{\nu} := \frac{d\nu(x,t)}{dt}.$$

Let $\mathcal{W}$ be the space of pairs $\vartheta := (y, \nu)$, a generic element of its tangent space $T_\vartheta \mathcal{W}$ at ϑ is then given by the pair $(\dot{y}, \dot{\nu}) = \dot{\vartheta}$. Moreover, the symbol $\mathfrak{P}(\mathcal{B}_0)$ denotes the algebra of parts of $\mathcal{B}_0$ that is the set of all subsets of $\mathcal{B}_0$ with non-vanishing volume measure and the same regularity properties of $\mathcal{B}_0$ itself.

Any *power* is such a map $\mathcal{P} : \mathfrak{P}(\mathcal{B}_0) \times T\mathcal{W} \to \mathbb{R}^+$ that

- $\mathcal{P}\left(\cdot, \vartheta, \dot{\vartheta}\right)$ is additive on disjoint parts and
- $\mathcal{P}(\mathfrak{b}, \vartheta, \cdot)$ is linear.

Of course, a key problem is the explicit representation of the power, that implies, in a sense, the selection of the type of interactions one is considering. In a wide sense, the explicit selection of the power is 'constitutive' because one selects the *type* of interactions occurring within a body. Here, once a part $\mathfrak{b}$ of $\mathcal{B}_0$ is selected arbitrarily, the attention is focused on the *external power* that is the power of all actions exchanged by $\mathfrak{b}$ with the rest of $\mathcal{B}_0$ and the external environment. As usual in continuum mechanics, I consider bulk and contact actions, the latter exerted through the boundary $\partial \mathfrak{b}$. No peculiar external bulk actions act directly on granules except the gravitation, because granules are only massive inert objects. Then the vector b indicates the sole standard macroscopic bulk force which includes, when relevant, inertial and non-inertial terms. At the boundary $\partial \mathfrak{b}$, standard contact interactions due to the relative change of place of neighboring material elements, are measured by means of the first Piola–Kirchhoff stress P while contact interactions due to relative changes of grain distributions *inside* material elements in contact through $\partial \mathfrak{b}$ are measured by a microstress $\mathcal{S}$.

By indicating by $\mathcal{P}_\mathfrak{b}^{ext}(\dot{y}, \dot{\nu})$ the *external power* over $\mathfrak{b}$ along (y, ν), with the pair $(\dot{y}, \dot{\nu}) := (\dot{y}(x,t), \dot{\nu}(x,t))$ belonging to the relevant $T_{(y,\nu)}\mathcal{W}$ at each (x, t), its explicit representation is then given by

$$\mathcal{P}_\mathfrak{b}^{ext}(\dot{y}, \dot{\nu}) := \int_\mathfrak{b} b \cdot \dot{y} dx + \int_{\partial \mathfrak{b}} (Pn \cdot \dot{y} + \mathcal{S}n \cdot \dot{\nu}) d\mathcal{H}^2,$$

where $d\mathcal{H}^2$ is the two-dimensional measure, n the normal to $\partial\mathfrak{b}$ in all places in which it is defined, that is everywhere except a finite number of corners and edges.

The basic interest here is on consequences of the requirement that $\mathcal{P}_{\mathfrak{b}}^{ext}$ be invariant under synchronous semi-classical changes in observers. Changes in synchronous classical observers, the class of changes in observers commonly used in standard continuum mechanics, are defined by (i) invariance of the time scale (from which the qualifier 'synchronous') and the reference place $\mathcal{B}_0$, (ii) isometric changes of the ambient space $\mathbb{R}^3$; they are thus governed by the action of the Euclidean group $\mathbb{R}^3 \ltimes SO(3)$ on $\mathbb{R}^3$. In this case one would have to consider the velocity $\dot{y}$ evaluated at x and t by the observer $\mathcal{O}$ and the one, indicated by $\dot{y}^{\#}$, measured by the observer $\mathcal{O}'$, shifting in time isometrically with respect to $\mathcal{O}$. Then one should pull-back in the frame $\mathcal{O}$ the rate $\dot{y}^{\#}$ obtaining a value $\dot{y}^* := Q^T \dot{y}^{\#}$ given by

$$\dot{y}^* = \dot{y} + c(t) + q(t) \times (x - x_0),\tag{6}$$

where $c(t)$ and $q(t)$ are respectively the relative translational and rotational velocities of the two observers. Basically, equation (6) means that $\mathcal{O}$ registers a velocity $\dot{y}^*$ measured by $\mathcal{O}'$, a velocity given by the rate $\dot{y}$ evaluated by $\mathcal{O}$ itself, augmented by the relative rigid motion $c(t) + q(t) \times (x - x_0)$ between $\mathcal{O}$ and $\mathcal{O}'$.

Contrary to the standard issue, here for the description of the shape and the motion of a granular body the manifold $\mathcal{M}$ is involved besides the interval of time, the reference place $\mathcal{B}_0$ and $\mathbb{R}^3$. As a consequence, since an *observer is a representation of all the geometrical environments necessary to describe the morphology and the motion of a body*, it is necessary to consider even the changes in the representation of $\mathcal{M}$ due to changes (here isometric) in observers. For this reason here changes in observers are called semi-classical. In fact, since the selection of $\mathcal{M}$ is in essence the choice of a model (a coarse grained model) of the real material texture, a change in observe in the ambient space 'alters' the perception of the material texture, that is the representation of $\mathcal{M}$.

Let then $t \mapsto Q(t) \in SO(3)$ be a smooth curve over $SO(3)$. From the definition of ν it follows that, after the action of the generic $Q(t) \in SO(3)$, it changes as

$$\nu^{\#} = Q^* \nu Q,$$

then

$$\dot{\nu}^{\#} = \dot{Q}^* \nu Q + Q^* \dot{\nu} Q + Q^* \nu \dot{Q},$$

the superposed dot meaning the derivative with respect to t.

By indicating by $\dot{\nu}^* := Q\dot{\nu}^{\#}Q^T$ the pull-back of $\dot{\nu}^{\#}$ by means of Q, it follows that

$$\dot{\nu}^* = \dot{\nu} + W\nu + \nu W = \dot{\nu} + \mathcal{A}q,$$

where $W \in so\,(3)$ for any t, q the axial vector of W and

$$\mathcal{A}\,(\nu) \in Hom\,\left(\mathbb{R}^3, T_\nu \mathcal{M}\right).$$

Specifically, $\mathcal{A}$ is given by the difference $e\nu - \nu e$, where e is Ricci's alternating tensor, namely

$$\mathcal{A}_{\alpha\beta i} = \left(e_{\alpha i \gamma} \nu_\beta^\gamma\right)^t - \nu_\alpha^\gamma e_{\gamma\beta i},$$

the exponent t indicating minor right transposition. Greek indices denote coordinates on $\mathcal{M}$ while Latin indices the ones on $\mathbb{R}^3$.

Note that the transformation leading to $\dot{y}^*$ and $\dot{\nu}^*$ are considered synchronous.

The requirement of invariance of the external power under the isometric changes in synchronous semi-classical observers described above is the basic axiom used in this section (see [17], [20] for its use in the case of general complex bodies).

Axiom of invariance. *At equilibrium the power of external actions is invariant under semi-classical changes in observers, namely*

$$\mathcal{P}_{\mathfrak{b}}^{ext}\,(\dot{y}, \dot{\nu}) = \mathcal{P}_{\mathfrak{b}}^{ext}\,(\dot{y}^*, \dot{\nu}^*)$$

for any choice of $\mathfrak{b}$, c *and* q.

The axiom is strictly equivalent to requiring

$$\mathcal{P}_{\mathfrak{b}}^{ext}\,(c + q \times x, \mathcal{A}q) = 0,$$

for any choice of $\mathfrak{b}$, c and q, a result which is the weak integral balance of actions. The arbitrariness of c, q and $\mathfrak{b}$ then implies the result below.

Theorem 1. *(i) If for any* $\mathfrak{b}$ *the vector fields* $x \mapsto Pn$ *and* $x \mapsto \mathcal{A}^* Sn$ *are defined over* $\partial\mathfrak{b}$ *and are integrable there, the integral balances of actions on* $\mathfrak{b}$ *hold:*

$$\int_{\mathfrak{b}} b \, dx + \int_{\partial\mathfrak{b}} Pn \, d\mathcal{H}^2 = 0, \tag{7}$$

$$\int_{\mathfrak{b}} \left((x - x_0) \times b + \mathcal{A}^*\beta\right) \, dx + \int_{\partial\mathfrak{b}} \left((x - x_0) \times Pn + \mathcal{A}^* Sn\right) \, d\mathcal{H}^2 = 0. \tag{8}$$

(ii) Moreover, if the tensor fields $x \mapsto P$ *and* $x \mapsto S$ *are of class* $C^1\,(\mathcal{B}_0) \cap C^0\,(\bar{\mathcal{B}}_0)$ *then*

$$Div P + b = 0 \tag{9}$$

and there exist a covector field $x \mapsto z \in T_{\nu(x)}\mathcal{M}$ *such that*

$$skw\,(PF^*) = e\,(\mathcal{A}^* z + (D\mathcal{A}^*)\,S) \tag{10}$$

and

$$DivS - z = 0, \tag{11}$$

with $z = z_1 + z_2$, $z_2 \in Ker\mathcal{A}^*$ *and* e *Ricci's tensor. (iii) Finally, if in addition the fields* $x \mapsto \dot{y}(x,t)$ *and* $x \mapsto \dot{\nu}(x,t)$ *are* $C^1(\mathcal{B}_0) \cap C^0(\bar{\mathcal{B}}_0)$ *then*

$$\mathcal{P}_{\mathfrak{b}}^{ext}(\dot{y}, \dot{\nu}) = \int_{\mathfrak{b}} \left(P \cdot \dot{F} + z \cdot \dot{\nu} + \mathcal{S} \cdot \dot{N} \right) \, dx. \tag{12}$$

Eulerian versions of the balance equations above can be obtained as usual by means of the inverse Piola transform. They read

$$div \, \sigma + b_a = 0,$$

$$skw(\sigma) = \mathsf{e}\left(\mathcal{A}^* z_a + (D\mathcal{A}^*)\, \mathcal{S}_a\right),$$

$$div\mathcal{S}_a - z_a = 0,$$

where the operator *div* is calculated with respect to y and

$$\sigma := (\det F)^{-1} P \left(F^{-1}\right)^*, \qquad b_a := (\det F)^{-1} b,$$

$$\mathcal{S}_a := (\det F)^{-1} \mathcal{S} \left(F^{-1}\right)^*, \qquad z_a := (\det F)^{-1} z$$

are the actual measures of external and internal actions.

A micromechanical interpretation of them can be done by exploiting the space $\hat{\mathbb{R}}^3$ where one imagines as above that the generic material element $\mathfrak{e}$ occupies a bounded compact domain with finite volume, the smallest domain containing the granules within $\mathfrak{e}$ in a compacted configuration. Other material elements can be also imagined to be expanded in $\hat{\mathbb{R}}^3$ and occupy compact domains having with the one just described only parts of the boundary in common. In what follows $\mathfrak{I}_\mathfrak{e}$ indicates the set indexing the granules within the material element $\mathfrak{e}$ (that are the granules not touching the boundary of $\mathfrak{e}$), while $\mathfrak{I}_{\partial\mathfrak{e}}$ is the set indexing the granules on the boundary of $\mathfrak{e}$. For $j \in \mathfrak{I}_\mathfrak{e}$, d_j is the diameter of the j-th granule within $\mathfrak{e}$. Let also $f_{(ij)}$ be the *total force* exchanged through contact between the granule i and the granule j, and $r_{(ij)}$ be the vector defined by $\hat{x}_{(i)} - \hat{x}_{(j)}$ that is the difference between the points occupied by the centres of mass of the granule i and the granule j. If two granules are in contact, then $\left|r_{(ij)}\right| \leq \frac{1}{2}(d_i + d_j)$. The total force $f_{(ij)}$ is considered applied to a contact point $\hat{x}_{(ij)}^c$ and $r_{(ij)}^c := \hat{x}_{(ij)}^c - \hat{x}$ is the vector connecting the centre of mass $\hat{x}$ of $\mathfrak{e}$ with $\hat{x}_{(ij)}^c$. Another total force $g_{(h)}$ is exchanged between $\mathfrak{e}$ and the h-th neighboring material element in contact with it. Such a force is applied at a point $\hat{x}_{(h)}^c$ and $r_{(h)} := \hat{x}_{(h)}^c - \hat{x}$ is the vector connecting the centre of mass $\hat{x}$ of $\mathfrak{e}$ with $\hat{x}_{(h)}^c$. With these premises, by making use of the isomorphism between $\mathbb{R}^3$ and $\hat{\mathbb{R}}^3$, I suggest the following microscopic interpretation of the coarse grained measures of interaction in Eulerian description:

$$z_a^{(\beta)}\left(\iota\left(\hat{x}\right)\right) = \frac{1}{|\mathfrak{e}|} \sum_{i \neq j \in \mathfrak{I}_\mathfrak{e}} f_{(ij)} \otimes r_{(ij)}^c, \tag{13}$$

$$\sigma^{(\beta)}\left(\iota\left(\hat{x}\right)\right) = \frac{1}{|\mathfrak{e}|} \sum_h g_{(h)} \otimes r_{(h)}, \tag{14}$$

$$S_a^{(\beta)}\left(\iota\left(\hat{x}\right)\right) = \frac{1}{|\mathfrak{e}|} \sum_h \sum_{i \in \mathfrak{I}_{\partial\mathfrak{e}}, \, j \in \mathfrak{I}_{\partial\mathfrak{e}_h}} f_{(ij)} \otimes r_{(ij)}^c \otimes r_{(ij)}^{c(h)}, \tag{15}$$

where $\iota\left(\hat{x}\right)$ corresponds to a generic point y in $\mathcal{B}$, and $r_{(ij)}^{c(h)} := \hat{x}_{(h)}^c - \hat{x}_{(h)}$ with $\hat{x}_{(h)}$ the centre of mass of the h-th material element $\mathfrak{e}_h$ in contact with $\mathfrak{e}$. The apex β indicates that $z_a^{(\beta)}$, $\sigma^{(\beta)}$, and $S_a^{(\beta)}$ are referred to a given specific β-th (here deterministic) configuration of granules. In a coarse grained sense one should then calculate z_a, σ, and S_a as averages aver all possible β's.

The interpretation above has strict analogies with the one of Edwards and Grinev [9] on Cauchy tensor in granular agglomeration. Differences are also evident: in [9] the generic material element coincides with only one granule so that the Cauchy stress can be identified basically with (14) (cf. formula (7) in [9]). However, in [9], z_a and S_a do not exist even if a fourth-rank hyperstress is introduced (see formula (17) in [9]) as a perturbation of the standard stress. Really, even here the microstress S_a can be considered as a hyperstress. This circumstance occurs when one imposes the internal constraint $\nu = F$ or $\nu = f(F)$, with f a $\mathcal{M}$-valued function. In this case the substructure becomes 'latent' in the sense of Capriz [4], the balance of substructural actions, namely (11) disappears and falls within Cauchy balance of standard forces generating a perturbation of the standard stress: the scheme reduces to the one of a second-grade material. However, contrary to [9], the hyperstress generated by the internal constraint linking ν to F is a third-rank tensor. Formulas (13)–(15) can be written of course in terms of the distribution function $\hat{\theta}^c$ of the contact points $\hat{x}^c$ and the distribution $\hat{\theta}^h$ of the centres of mass $\hat{x}^h$ of the material elements in a neighborhood of $\mathfrak{e}$, the latter distribution necessary for the microstress S_a, to account for uncertainties about the exact location of contact points and neighboring material elements. However, even acting so, the physical interpretation of the measures of interactions does not change.

4 Evolution of the Local Numerosity of Granules

When granules migrate from a material element into another, the numerosity satisfies the continuity equation

$$\frac{d}{dt} \int_\mathfrak{b} \alpha \, dx + \int_{\partial\mathfrak{b}} \omega \cdot n \, d\mathcal{H}^2 = 0$$

for any $\mathfrak{b} \in \mathfrak{P}(\mathcal{B}_0)$. In this equation, the vector ω is the flux through the boundary $\partial\mathfrak{b}$, n the outward unit normal to regular points of $\partial\mathfrak{b}$. If $(x,t) \mapsto \alpha(x,t)$ and $(x,t) \mapsto \omega(x,t)$ are C^1 maps, the arbitrariness of $\mathfrak{b}$ implies

$$\dot{\alpha} + Div\, \omega = 0, \tag{16}$$

a continuity equation which coincides with the conservation of mass when

$$\int_{\partial\mathcal{B}_0} \omega \cdot n \; d\mathcal{H}^2 = 0,$$

since it has been presumed that all granules have equal mass.

Migration of granules implies also the *growth of local configurational entropy*, entropy related with the loss of information about the distribution of granules within the generic material element suffering migration. For any $\mathfrak{b} \in \mathfrak{P}(\mathcal{B}_0)$, the entropy production due to the migration of granules across the boundary $\partial\mathfrak{b}$ is presumed to be given by

$$\int_{\partial\mathfrak{b}} h \cdot n \; d\mathcal{H}^2$$

where h is a $C^1\left(\mathcal{B}_0 \times [0, \bar{t}], \mathbb{R}^3\right)$ vector density along $\partial\mathfrak{b}$ (this assumption has been made also in [13] in the case of mass transport of one specie in multiphase materials and in [18] in the case of migration of general substructures in complex fluids, a paper, the latter, in which the interpretation of h as *configurational* entropy has been proposed). The vector flux h is assumed to be proportional to the flux of granules, namely

$$h = \mu\omega. \tag{17}$$

The map $(x,t) \mapsto \mu(x,t)$, $\mu \in C^1\left(\mathcal{B}_0 \times [0, \bar{t}]\right)$, is properly the *chemical potential* since all granules have the same mass. Of course μ can be defined because the assembly of granules is considered as a single continuum.

Let $\Psi(\mathfrak{b}, t)$ be the *free energy of* $\mathfrak{b} \in \mathfrak{P}(\mathcal{B}_0)$ at the instant t. By taking into account the contribution of the configurational entropy due to the migration of substructures, the isothermal version of the Clausius–Duhem inequality (then a mechanical dissipation inequality) *along* (y, ν) can be then written as

$$\frac{d}{dt}\Psi(\mathfrak{b}, t) - \int_{\partial\mathfrak{b}} h \cdot n \; d\mathcal{H}^2 - \mathcal{P}_\mathfrak{b}^{ext}(\dot{y}, \dot{\nu}) \leq 0,$$

for any $\mathfrak{b} \in \mathfrak{P}(\mathcal{B}_0)$ and any choice of the rates involved. A standard assumption in continuum mechanics is that the free energy of any $\mathfrak{b}$ is given by

$$\Psi(\mathfrak{b}, t) := \int_\mathfrak{b} \psi \; dx,$$

with ψ a C^1 density. The arbitrariness of $\mathfrak{b}$ and Theorem 1 imply the local dissipation inequality

$$\dot{\psi} - Div\, h - P \cdot \dot{F} + z \cdot \dot{\nu} + \mathcal{S} \cdot \dot{N} \leq 0. \tag{18}$$

In addition to (17), constitutive assumptions about the free energy and the representatives of interactions have to be accounted for in order to finding consequences of (18). Constitutive assumptions then are

$$\psi := \psi\left(F, \nu, N, \mu, D\mu\right), \tag{19}$$

$$P := P\left(F, \nu, N, \mu\right), \tag{20}$$

$$z := z\left(F, \nu, N, \mu\right), \tag{21}$$

$$\mathcal{S} := \mathcal{S}\left(F, \nu, N, \mu\right). \tag{22}$$

By inserting (17) and (19)–(22) in (18) and making use of (16), the arbitrariness of the rates involved implies the results listed in what follows.

- Under the assumptions above the free energy density ψ *cannot* depend on $D\mu$.
- The constitutive restrictions listed below follow:

$$P = \partial_F \psi\left(F, \nu, N, \mu\right), \quad \mathcal{S} = \partial_N \psi\left(F, \nu, N, \mu\right), \tag{23}$$

$$z = \mu D_\nu \alpha + \partial_\nu \psi\left(F, \nu, N, \mu\right). \tag{24}$$

- The reduced dissipation inequality

$$\omega \cdot D\mu \geq 0 \tag{25}$$

implies

$$\omega = AD\mu, \tag{26}$$

with A the *mobility* tensor, a second-rank definite positive tensor.

From (24) one gets immediately

$$\mu = \frac{1}{|D_\nu \alpha|^2}\, \langle z - \partial_\nu \psi, D_\nu \alpha \rangle_{T_\nu^* \mathcal{M}}, \tag{27}$$

where $\langle \cdot, \cdot \rangle_{T_\nu^* \mathcal{M}}$ is a scalar product over $T_\nu^* \mathcal{M}$. By using (26) and (11), one changes (16) as

$$\dot{\alpha} = -Div\left(AD\left(\frac{1}{|D_\nu \alpha|^2}\, \langle Div\mathcal{S} - \partial_\nu \psi, D_\nu \alpha \rangle_{T_\nu^* \mathcal{M}}\right)\right), \tag{28}$$

for sufficient smoothness of the fields involved. Equation (28) is the desired evolution equation of the numerosity. A distributional version of it holds also.

Once a granular body is at rest, re-compaction may change its mass density up to 20% so that (28) describes effects of clustering of granules that are often non-negligible. Equation (28) is a special case in Lagrangian representation of the evolution equation for general substructures derived in [18]. It suggests that the basic mechanism ruling the migration of granules is the competition of substructural actions between neighboring material elements. Such an interpretation follows from the presence of the gradient of the projection of $DivS - \partial_\nu\psi \in T_\nu^*\mathcal{M}$ along $D_\nu\alpha$. Due to the interpretation of the measures of interaction in terms of contact forces between the granules within the generic material element, in the case in which all granules have equal shape, from (28) one may infer that the granules migrate between a certain material element and the neighboring elements in with a less number of contacts occur. The interpretation is suggested by the presence of the right-hand-side term. It is more evident in the case of *latence*. In this case, by selecting for the sake of simplicity the internal constraint as $\nu = F$, one gets from (28)

$$\dot{\alpha} = -Div\left(AD\left(\frac{1}{|D_F\alpha|^2}\langle Div\partial_{DF}\psi - \partial_F\psi, D_F\alpha\rangle\right)\right), \qquad (29)$$

where now the scalar product $\langle\cdot,\cdot\rangle$ in (29) is on $T_F^*Hom\left(T_x\mathcal{B}_0, T_{y(x)}\mathcal{B}\right)$. The difference $Div\partial_{DF}\psi - \partial_F\psi$ is the difference between the first Piola–Kirchhoff stress and the perturbation induced on it by the hyperstress $\partial_{DF}\psi$. The discrete evolution in time of (29) may clarify the interpretation: One gets in fact

$$\alpha_{i+1} = \alpha_i - (\Delta t_i)Div\left(AD\left(\frac{1}{|D_F\alpha_i|^2}\langle Div\partial_{DF}\psi - \partial_F\psi, D_F\alpha_i\rangle\right)\right), \quad (30)$$

where Δt_i is the time step at t_i. So, since α_0 is assigned at the initial time as a constitutive prescription, the subsequent steps in time are determined by the actual values of the stress and the hyperstress. No evolution occurs when $\langle Div\partial_{DF}\psi - \partial_F\psi, D_F\alpha_i\rangle = 0$ or, more generally, when $\langle DivS - \partial_\nu\psi, D_\nu\alpha\rangle_{T_\nu^*\mathcal{M}} = 0$.

5 A Single Granule Coinciding with the Generic Material Element

The generic material element can be also considered as composed by a single granule only. The multifield setting presented above applies when one aims to account for local rotations of granules. In this case ν can be 'identified' with $\hat{R}$, namely ν can be considered as an element of $SO(3)$. However, one could reduce the tensor order of ν by imagining it as a unit vector belonging to the unit sphere S^2 in $\mathbb{R}^3$. The isomorphism between S^2 and $SO(3)$ assures this possibility. The scheme becomes then the one of continua with spin structure, a scheme adopted below.

Migration of granules cannot be treated as above because the material element is no more a family of granules that may loose or get fellows. However, the motion of a single granule relative to the rest of the body can be described directly by considering it as a point defect. I review here some results collected in [19] and valid for continua with spin structure. They apply to the case analyzed in this section.

Imagine a granule located at $\bar{x}$ in $\mathcal{B}_0$, a granule that moves in $\mathcal{B}$ relatively to $\mathcal{B}$ itself. By means of y^{-1} one may picture this motion in $\mathcal{B}_0$ in a sort of non-material motion described by a differentiable map $t \mapsto \bar{x}(t)$, a motion with (material) velocity $w := \frac{d\bar{x}(t)}{dt}$. A driving force $\mathfrak{f}$ is power conjugated with w. Since the motion $t \mapsto \bar{x}(t)$ is 'virtual' (in the sense that it is the 'shadow' in $\mathcal{B}_0$ of $t \mapsto y(\bar{x}, t)$, that is the actual motion of the granule in $\mathcal{B}$ relatively to $\mathcal{B}$ itself), the force $\mathfrak{f}$ is 'virtual' too so it has to be expressed in terms of the true standard and substructural actions, the ones that push eventually the granule to move in $\mathcal{B}$ relatively to the neighboring fellows. Let $\mathfrak{b}_r$ be a sphere of radius r contained in $\mathcal{B}_0$ and centered at $\bar{x}$ at any t. At each t the driving force $\mathfrak{f}$ is given by

$$\mathfrak{f} = \lim_{r \to 0} \int_{\partial \mathfrak{b}_r} \mathbb{P}n \, d\mathcal{H}^2, \tag{31}$$

where $\mathbb{P}$ is the extended Hamilton–Eshelby tensor given by $\mathbb{P} := \psi I - F^* P - N^* S$ (see [19] for the proof of (31) in a wider setting). At each t, local balances of standard and substructural actions hold also at $\bar{x}$. They read

$$\lim_{r \to 0} \int_{\partial \mathfrak{b}_r} Pn \, d\mathcal{H}^2 = 0, \qquad \lim_{r \to 0} \int_{\partial \mathfrak{b}_r} Sn \, d\mathcal{H}^2 = 0$$

(see also [19] for the proof).

The driving force $\mathfrak{f}$ is intrinsically dissipative in the sense that $\mathfrak{f} \cdot w \geq 0$ with the identity holding when $w \neq 0$. As a consequence $\mathfrak{f} = g(w) w$, with $g(w)$ a definite positive scalar function that has constitutive nature. The velocity w is different from zero only when $\mathfrak{f}$ satisfies a certain threshold along a given direction. In this case the balance (31) becomes the evolution equation

$$g(w) w = \lim_{r \to 0} \int_{\partial \mathfrak{b}_r} \mathbb{P}n \, d\mathcal{H}^2.$$

Since w can be written as $|w| s$, $s \in S^2$, for prominent anisotropy of the body, the 'strength' around $\bar{x}$ can be described by making use of a map $f = S^2 \to \mathbb{R}^+$. In this case $\mathfrak{f}$ is called *subcritical* when $\mathfrak{f} \cdot s < f(s)$ for all $s \in S^2$, *critical* when there exist some $s \in S^2$ such that $\mathfrak{f} \cdot s = f(s)$ while subcriticality is granted for all directions, *supercritical* when there exist some $s \in S^2$ such that $\mathfrak{f} \cdot s > f(s)$. The dissipation $\mathfrak{D}$ along the motion $t \mapsto \bar{x}(t)$ is given by $\mathfrak{D}(\mathfrak{f}, s) = (\mathfrak{f} \cdot s) |w| = g(w) |w|^2$ and occurs along directions where $\mathfrak{f}$

is supercritical. When many directions assure supercriticality, the maximum dissipation principle indicates the direction s along which the motion develops, namely s is the argument of

$$\max_{s \in S^2} \{ \mathfrak{D}\,(\mathfrak{f}, s) \ \text{s. t.} \ \mathfrak{f} \cdot s > f\,(s) \}.$$

Acknowledgement

I wish to thank Gianfranco Capriz for many essential discussions on the nature of granular matter. My gratitude goes also to Serena Poppi who pushed indirectly me to conclude this chapter during contemporary pressing work. The support of the Italian National Group of Mathematical Physics (GNFM-INDAM) and of MIUR under the grant 2005085973 − *"Resistenza e degrado di interfacce in materiali e strutture"* − COFIN 2005 is acknowledged.

References

1. Aronson, I.S., Tsimring, L.S.: Patterns and collective behavior in granular media: theoretical concepts. Rev. Modern Phys., **78**, 641–692 (2006)
2. Bobylev, A.V., Carrillo, J.A., Gamba, I.M.: On some properties of kinetic and hydrodynamic equations for inelastic interactions. J. Stat. Phys., **98**, 743–773 (2000)
3. Bobylev, A., Cercignani, C., Toscani, G.: Proof of an asymptotic property of the Boltzmann for granular materials. J. Stat. Phys., **111**, 403–416 (2003)
4. Capriz, G.: Continua with latent microstructure. Arch. Rational Mech. Anal., **90**, 43–56 (1985)
5. Capriz, G.: Continua with Microstructure. Springer Tracts in Natural Philosophy. Springer-Verlag, Berlin Heidelberg New York, **35** (1989)
6. Capriz, G.: Elementary preamble to a theory of granular gases. Rend. Sem. Mat. Univ. Padova, **110**, 179–198 (2003)
7. Capriz, G.: Pseudofluids. In: Capriz, G., Mariano, P.M. (eds) Material Substructures in Complex Bodies: from Atomic Level to Continuum. Elsevier Science B.V., Amsterdam, 238–261 (2007)
8. Cercignani, C.: Microscopic foundations of the mechanics of gases and granular materials. In: Capriz, G., Mariano, P.M. (eds) Material Substructures in Complex Bodies: from Atomic Level to Continuum. Elsevier Science B.V., Amsterdam, 63–79 (2007)
9. Edwards, S.F., Grinev, D.V.: The statistical–mechanical theory of stress transmission in granular matter. Physica A, **263**, 545–553 (1999)
10. Ericksen, J.L.: Kinematics of macromolecules. Arch. Rational Mech. Anal., **9**, 1–8 (1962)
11. Gamba, I.M., Panferov, V., Villani, C.: On the Boltzmann equation for diffusively excited granular media. Comm. Math. Phys., **246**, 503–541 (2004)
12. Goddard, J.D.: Material instability in comlex fluids. Ann. Rev. Fluid Mech., **35**, 113–133 (2003)

13. Gurtin, M.E.: Generalized Ginzburg–Landau and Cahn–Hilliard equations based on a microforce balance. Physica D, **92**, 178–192 (1996)
14. Jaeger, H.M., Nagel, S.R., Behringer, R.P.: Granular solids, liquid and gases. Rev. Modern Phys., **68**, 1259–1273 (1996)
15. Kadanoff, L.P.: Built upon sand: Theoretical ideas inspired by granular flows. Rev. Modern Phys., **71**, 435–444 (1999)
16. Krimer, D.O., Pfitzener, M., Bräuer, K., Jiang, Y., Liu, M.: Granular elasticity: General considerations and the stress dip in sand piles. Phys. Rev. E, **74**, 061310-1-10 (2006)
17. Mariano, P.M.: Multifield theories in mechanics of solids. Adv. Appl. Mech., **38**, 1–93 (2002)
18. Mariano, P.M.: Migration of substructures in complex fluids. J. Phys. A, **38**, 6823–6839 (2005)
19. Mariano, P.M.: Continua with spin structure. In: Capriz, G., Mariano, P.M. (eds) Material Substructures in Complex Bodies: from Atomic Level to Continuum. Elsevier Science B.V., Amsterdam, 314–334 (2007)
20. Mariano, P.M., Stazi, F.L.: Computational aspects of the mechanics of complex bodies. Arch. Comp. Meth. Eng., **12**, 391–478 (2005)
21. Mitarai, N., Hayakawa, H., Nakanishi, H.: Collisional granular flow as a micropolar fluid. Phys. Rev. Lett., **88**, 174301-1-4 (2002)
22. Šilhavý, M.: The Mechanics and Thermodynamics of Continuous Media. Springer–Verlag, Berlin (1997)
23. Truesdell, C.A., Noll, W.: The Non-Linear Field Theories of Mechanics. Third edition, Springer–Verlag, Berlin (2004)

Index

Lecture Notes in Mathematics

For information about earlier volumes
please contact your bookseller or Springer
LNM Online archive: springerlink.com

Vol. 1854: O. Saeki, Topology of Singular Fibers of Differential Maps (2004)

Vol. 1855: G. Da Prato, P.C. Kunstmann, I. Lasiecka, A. Lunardi, R. Schnaubelt, L. Weis, Functional Analytic Methods for Evolution Equations. Editors: M. Iannelli, R. Nagel, S. Piazzera (2004)

Vol. 1856: K. Back, T.R. Bielecki, C. Hipp, S. Peng, W. Schachermayer, Stochastic Methods in Finance, Bressanone/Brixen, Italy, 2003. Editors: M. Fritelli, W. Runggaldier (2004)

Vol. 1857: M. Émery, M. Ledoux, M. Yor (Eds.), Séminaire de Probabilités XXXVIII (2005)

Vol. 1858: A.S. Cherny, H.-J. Engelbert, Singular Stochastic Differential Equations (2005)

Vol. 1859: E. Letellier, Fourier Transforms of Invariant Functions on Finite Reductive Lie Algebras (2005)

Vol. 1860: A. Borisyuk, G.B. Ermentrout, A. Friedman, D. Terman, Tutorials in Mathematical Biosciences I. Mathematical Neurosciences (2005)

Vol. 1861: G. Benettin, J. Henrard, S. Kuksin, Hamiltonian Dynamics – Theory and Applications, Cetraro, Italy, 1999. Editor: A. Giorgilli (2005)

Vol. 1862: B. Helffer, F. Nier, Hypoelliptic Estimates and Spectral Theory for Fokker-Planck Operators and Witten Laplacians (2005)

Vol. 1863: H. Führ, Abstract Harmonic Analysis of Continuous Wavelet Transforms (2005)

Vol. 1864: K. Efstathiou, Metamorphoses of Hamiltonian Systems with Symmetries (2005)

Vol. 1865: D. Applebaum, B.V. R. Bhat, J. Kustermans, J. M. Lindsay, Quantum Independent Increment Processes I. From Classical Probability to Quantum Stochastic Calculus. Editors: M. Schürmann, U. Franz (2005)

Vol. 1866: O.E. Barndorff-Nielsen, U. Franz, R. Gohm, B. Kümmerer, S. Thorbjønsen, Quantum Independent Increment Processes II. Structure of Quantum Lévy Processes, Classical Probability, and Physics. Editors: M. Schürmann, U. Franz, (2005)

Vol. 1867: J. Sneyd (Ed.), Tutorials in Mathematical Biosciences II. Mathematical Modeling of Calcium Dynamics and Signal Transduction. (2005)

Vol. 1868: J. Jorgenson, S. Lang, $Pos_n(R)$ and Eisenstein Series. (2005)

Vol. 1869: A. Dembo, T. Funaki, Lectures on Probability Theory and Statistics. Ecole d'Eté de Probabilités de Saint-Flour XXXIII-2003. Editor: J. Picard (2005)

Vol. 1870: V.I. Gurariy, W. Lusky, Geometry of Müntz Spaces and Related Questions. (2005)

Vol. 1871: P. Constantin, G. Gallavotti, A.V. Kazhikhov, Y. Meyer, S. Ukai, Mathematical Foundation of Turbulent Viscous Flows, Martina Franca, Italy, 2003. Editors: M. Cannone, T. Miyakawa (2006)

Vol. 1872: A. Friedman (Ed.), Tutorials in Mathematical Biosciences III. Cell Cycle, Proliferation, and Cancer (2006)

Vol. 1873: R. Mansuy, M. Yor, Random Times and Enlargements of Filtrations in a Brownian Setting (2006)

Vol. 1874: M. Yor, M. Émery (Eds.), In Memoriam Paul-André Meyer - Séminaire de Probabilités XXXIX (2006)

Vol. 1875: J. Pitman, Combinatorial Stochastic Processes. Ecole d'Eté de Probabilités de Saint-Flour XXXII-2002. Editor: J. Picard (2006)

Vol. 1876: H. Herrlich, Axiom of Choice (2006)

Vol. 1877: J. Steuding, Value Distributions of L-Functions (2007)

Vol. 1878: R. Cerf, The Wulff Crystal in Ising and Percolation Models, Ecole d'Eté de Probabilités de Saint-Flour XXXIV-2004. Editor: Jean Picard (2006)

Vol. 1879: G. Slade, The Lace Expansion and its Applications, Ecole d'Eté de Probabilités de Saint-Flour XXXIV-2004. Editor: Jean Picard (2006)

Vol. 1880: S. Attal, A. Joye, C.-A. Pillet, Open Quantum Systems I, The Hamiltonian Approach (2006)

Vol. 1881: S. Attal, A. Joye, C.-A. Pillet, Open Quantum Systems II, The Markovian Approach (2006)

Vol. 1882: S. Attal, A. Joye, C.-A. Pillet, Open Quantum Systems III, Recent Developments (2006)

Vol. 1883: W. Van Assche, F. Marcellàn (Eds.), Orthogonal Polynomials and Special Functions, Computation and Application (2006)

Vol. 1884: N. Hayashi, E.I. Kaikina, P.I. Naumkin, I.A. Shishmarev, Asymptotics for Dissipative Nonlinear Equations (2006)

Vol. 1885: A. Telcs, The Art of Random Walks (2006)

Vol. 1886: S. Takamura, Splitting Deformations of Degenerations of Complex Curves (2006)

Vol. 1887: K. Habermann, L. Habermann, Introduction to Symplectic Dirac Operators (2006)

Vol. 1888: J. van der Hoeven, Transseries and Real Differential Algebra (2006)

Vol. 1889: G. Osipenko, Dynamical Systems, Graphs, and Algorithms (2006)

Vol. 1890: M. Bunge, J. Funk, Singular Coverings of Toposes (2006)

Vol. 1891: J.B. Friedlander, D.R. Heath-Brown, H. Iwaniec, J. Kaczorowski, Analytic Number Theory, Cetraro, Italy, 2002. Editors: A. Perelli, C. Viola (2006)

Vol. 1892: A. Baddeley, I. Bárány, R. Schneider, W. Weil, Stochastic Geometry, Martina Franca, Italy, 2004. Editor: W. Weil (2007)

Vol. 1893: H. Hanßmann, Local and Semi-Local Bifurcations in Hamiltonian Dynamical Systems, Results and Examples (2007)

Vol. 1894: C.W. Groetsch, Stable Approximate Evaluation of Unbounded Operators (2007)

Vol. 1895: L. Molnár, Selected Preserver Problems on Algebraic Structures of Linear Operators and on Function Spaces (2007)

Vol. 1896: P. Massart, Concentration Inequalities and Model Selection, Ecole d'Été de Probabilités de Saint-Flour XXXIII-2003. Editor: J. Picard (2007)

Vol. 1897: R. Doney, Fluctuation Theory for Lévy Processes, Ecole d'Été de Probabilités de Saint-Flour XXXV-2005. Editor: J. Picard (2007)

Vol. 1898: H.R. Beyer, Beyond Partial Differential Equations, On linear and Quasi-Linear Abstract Hyperbolic Evolution Equations (2007)

Vol. 1899: Séminaire de Probabilités XL. Editors: C. Donati-Martin, M. Émery, A. Rouault, C. Stricker (2007)

Vol. 1900: E. Bolthausen, A. Bovier (Eds.), Spin Glasses (2007)

Vol. 1901: O. Wittenberg, Intersections de deux quadriques et pinceaux de courbes de genre 1, Intersections of Two Quadrics and Pencils of Curves of Genus 1 (2007)

Vol. 1902: A. Isaev, Lectures on the Automorphism Groups of Kobayashi-Hyperbolic Manifolds (2007)

Vol. 1903: G. Kresin, V. Maz'ya, Sharp Real-Part Theorems (2007)

Vol. 1904: P. Giesl, Construction of Global Lyapunov Functions Using Radial Basis Functions (2007)

Vol. 1905: C. Prévôt, M. Röckner, A Concise Course on Stochastic Partial Differential Equations (2007)

Vol. 1906: T. Schuster, The Method of Approximate Inverse: Theory and Applications (2007)

Vol. 1907: M. Rasmussen, Attractivity and Bifurcation for Nonautonomous Dynamical Systems (2007)

Vol. 1908: T.J. Lyons, M. Caruana, T. Lévy, Differential Equations Driven by Rough Paths, Ecole d'Été de Probabilités de Saint-Flour XXXIV-2004 (2007)

Vol. 1909: H. Akiyoshi, M. Sakuma, M. Wada, Y. Yamashita, Punctured Torus Groups and 2-Bridge Knot Groups (I) (2007)

Vol. 1910: V.D. Milman, G. Schechtman (Eds.), Geometric Aspects of Functional Analysis. Israel Seminar 2004-2005 (2007)

Vol. 1911: A. Bressan, D. Serre, M. Williams, K. Zumbrun, Hyperbolic Systems of Balance Laws. Cetraro, Italy 2003. Editor: P. Marcati (2007)

Vol. 1912: V. Berinde, Iterative Approximation of Fixed Points (2007)

Vol. 1913: J.E. Marsden, G. Misiołek, J.-P. Ortega, M. Perlmutter, T.S. Ratiu, Hamiltonian Reduction by Stages (2007)

Vol. 1914: G. Kutyniok, Affine Density in Wavelet Analysis (2007)

Vol. 1915: T. Bıyıkoğlu, J. Leydold, P.F. Stadler, Laplacian Eigenvectors of Graphs. Perron-Frobenius and Faber-Krahn Type Theorems (2007)

Vol. 1916: C. Villani, F. Rezakhanlou, Entropy Methods for the Boltzmann Equation. Editors: F. Golse, S. Olla (2008)

Vol. 1917: I. Veselić, Existence and Regularity Properties of the Integrated Density of States of Random Schrödinger (2008)

Vol. 1918: B. Roberts, R. Schmidt, Local Newforms for GSp(4) (2007)

Vol. 1919: R.A. Carmona, I. Ekeland, A. Kohatsu-Higa, J.-M. Lasry, P.-L. Lions, H. Pham, E. Taflin, Paris-Princeton Lectures on Mathematical Finance 2004. Editors: R.A. Carmona, E. Çinlar, I. Ekeland, E. Jouini, J.A. Scheinkman, N. Touzi (2007)

Vol. 1920: S.N. Evans, Probability and Real Trees. Ecole d'Été de Probabilités de Saint-Flour XXXV-2005 (2008)

Vol. 1921: J.P. Tian, Evolution Algebras and their Applications (2008)

Vol. 1922: A. Friedman (Ed.), Tutorials in Mathematical BioSciences IV. Evolution and Ecology (2008)

Vol. 1923: J.P.N. Bishwal, Parameter Estimation in Stochastic Differential Equations (2008)

Vol. 1924: M. Wilson, Littlewood-Paley Theory and Exponential-Square Integrability (2008)

Vol. 1925: M. du Sautoy, L. Woodward, Zeta Functions of Groups and Rings (2008)

Vol. 1926: L. Barreira, V. Claudia, Stability of Nonautonomous Differential Equations (2008)

Vol. 1927: L. Ambrosio, L. Caffarelli, M.G. Crandall, L.C. Evans, N. Fusco, Calculus of Variations and Non-Linear Partial Differential Equations. Cetraro, Italy 2005. Editors: B. Dacorogna, P. Marcellini (2008)

Vol. 1928: J. Jonsson, Simplicial Complexes of Graphs (2008)

Vol. 1929: Y. Mishura, Stochastic Calculus for Fractional Brownian Motion and Related Processes (2008)

Vol. 1930: J.M. Urbano, The Method of Intrinsic Scaling. A Systematic Approach to Regularity for Degenerate and Singular PDEs (2008)

Vol. 1931: M. Cowling, E. Frenkel, M. Kashiwara, A. Valette, D.A. Vogan, Jr., N.R. Wallach, Representation Theory and Complex Analysis. Venice, Italy 2004. Editors: E.C. Tarabusi, A. D'Agnolo, M. Picardello (2008)

Vol. 1932: A.A. Agrachev, A.S. Morse, E.D. Sontag, H.J. Sussmann, V.I. Utkin, Nonlinear and Optimal Control Theory. Cetraro, Italy 2004. Editors: P. Nistri, G. Stefani (2008)

Vol. 1933: M. Petkovic, Point Estimation of Root Finding Methods (2008)

Vol. 1934: C. Donati-Martin, M. Émery, A. Rouault, C. Stricker (Eds.), Séminaire de Probabilités XLI (2008)

Vol. 1935: A. Unterberger, Alternative Pseudodifferential Analysis (2008)

Vol. 1936: P. Magal, S. Ruan (Eds.), Structured Population Models in Biology and Epidemiology (2008)

Vol. 1937: G. Capriz, P. Giovine, P.M. Mariano (Eds.), Mathematical Models of Granular Matter (2008)

Vol. 1938: D. Auroux, F. Catanese, M. Manetti, P. Seidel, B. Siebert, I. Smith, G. Tian, Symplectic 4-Manifolds and Algebraic Surfaces. Cetraro, Italy 2003. Editors: F. Catanese, G. Tian (2008)

Vol. 1939: D. Boffi, F. Brezzi, L. Demkowicz, R.G. Durán, R.S. Falk, M. Fortin, Mixed Finite Elements, Compatibility Conditions, and Applications. Cetraro, Italy 2006. Editors: D. Boffi, L. Gastaldi (2008)

Vol. 1940: J. Banasiak, V. Capasso, M.A.J. Chaplain, M. Lachowicz, J. Miękisz, Multiscale Problems in the Life Sciences. From Microscopic to Macroscopic. Będlewo, Poland 2006. Editors: V. Capasso, M. Lachowicz (2008)

Vol. 1941: S.M.J. Haran, Arithmetical Investigations. Representation Theory, Orthogonal Polynomials, and Quantum Interpolations (2008)

Vol. 1942: S. Albeverio, F. Flandoli, Y.G. Sinai, SPDE in Hydrodynamic. Recent Progress and Prospects. Cetraro, Italy 2005. Editors: G. Da Prato, M. Röckner (2008)

Vol. 1943: L.L. Bonilla (Ed.), Inverse Problems and Imaging. Martina Franca, Italy 2002 (2008)

Vol. 1944: A. Di Bartolo, G. Falcone, P. Plaumann, K. Strambach, Algebraic Groups and Lie Groups with Few Factors (2008)

Vol. 1945: F. Brauer, P. van den Driessche, J. Wu (Eds.), Mathematical Epidemiology (2008)

Recent Reprints and New Editions

Vol. 1702: J. Ma, J. Yong, Forward-Backward Stochastic Differential Equations and their Applications. 1999 – Corr. 3rd printing (2007)

Vol. 830: J.A. Green, Polynomial Representations of GL_n, with an Appendix on Schensted Correspondence and Littelmann Paths by K. Erdmann, J.A. Green and M. Schoker 1980 – 2nd corr. and augmented edition (2007)

Vol. 1693: S. Simons, From Hahn-Banach to Monotonicity (Minimax and Monotonicity 1998) – 2nd exp. edition (2008)

Vol. 470: R.E. Bowen, Equilibrium States and the Ergodic Theory of Anosov Diffeomorphisms. With a preface by D. Ruelle. Edited by J.-R. Chazottes. 1975 – 2nd rev. edition (2008)

Vol. 523: S.A. Albeverio, R.J. Høegh-Krohn, S. Mazzucchi, Mathematical Theory of Feynman Path Integral. 1976 – 2nd corr. and enlarged edition (2008)